Bureau of Animal Industry

The animal parasites of sheep

Bureau of Animal Industry

The animal parasites of sheep

ISBN/EAN: 9783337238209

Printed in Europe, USA, Canada, Australia, Japan

Cover: Foto ©berggeist007 / pixelio.de

More available books at **www.hansebooks.com**

U. S. DEPARTMENT OF AGRICULTURE.

BUREAU OF ANIMAL INDUSTRY.

GOVERNMENT PRINTING OFFICE.
1890.

TABLE OF CONTENTS.

3

INDEX OF ILLUSTRATIONS.

LETTER OF TRANSMITTAL.

WASHINGTON, D. C., *April* 21, 1890.

SIR: I have the honor to submit herewith a report upon the parasites of sheep, which has been prepared with much care and will prove of permanent value to all owners of this class of our domesticated animals. The information heretofore attainable on this subject in the United States has been fragmentary and in many cases unreliable, although the parasitic diseases of sheep are among the most frequent and serious maladies by which this species of animals are affected.

It has been the aim in the preparation of this volume to make the descriptions and the illustrations so plain that any one will be able to identify the parasites which he may find in his flock, and yet the subject is in some of its aspects so technical that it could not be presented entirely in popular language. The technical descriptions which it is deemed necessary to insert have, however, been placed in small type, and those not interested in the characters by which the species are identified can omit such paragraphs. The symptoms and appearances presented by diseased animals and the treatment of the diseases have been given at considerable length, and these will be read with interest by all who desire information on this subject. The illustrations are a prominent feature of the work, having been drawn and lithographed with the greatest care, and every attention given to make them accurate in their most minute details. Nearly all of these are original and were drawn from nature.

The nodular disease of the intestines, together with its cause, is described for the first time in these pages. This disease is common and wide-spread, but its cause and nature were mysterious until they were discovered through the investigations of this Bureau. We have here once again a demonstration of the value of systematic, scientific investigation of the diseases of animals, for the results obtained by the study of this malady are among the most interesting contributions of modern research. The facts obtained in the investigations of the fringed tape-worm and the hair lung-worm are also of more than ordinary interest.

The subject of parasites and parasitic diseases is one of great importance, and must become more prominent as the number of domesticated animals in the country increases and the pastures become more

7

limited in comparison with the flocks which graze upon them. Under such conditions parasites multiply more rapidly, and their ravages become more alarming. For this reason the time has come when we must pay more attention to these organisms and study more assiduously the means of controlling them, if we would preserve that healthfulness and vigor for which the animals of this country have heretofore been noted. It is hoped that the systematic treatment of the subject presented in the accompanying volume may assist in accomplishing this object.

Very respectfully,

D. E. SALMON,
Chief of the Bureau of Animal Industry.

Hon. J. M. RUSK,
Secretary of Agriculture.

ANIMAL PARASITES OF SHEEP.

GENERAL REMARKS.

In 1782, Gœze, a distinguished German naturalist, wrote: "Among all mammals except the horse, the sheep appears to be most harassed by worms." He thus called attention at that early period to the great abundance of ovine parasites, an abundance which have transmitted their posterity in comparatively undiminished numbers.

The presentation of all the facts now known concerning these parasites, their structure, their life histories, the injuries they cause, and the methods of prevention and treatment, together with such new material as may have been learned concerning them, needs no apology to the sheep owner, for he is alive to the fact that the majority of his losses is due to these parasites. The sheep industry of the United States embraces the product of 42,599,079 sheep, valued at $90,640,369.* Dependent on these sheep and their products are an army of men and their families, from the flock-master and his help to the consumers of the flesh and the manufacturers of the fleece. Add to this the value of the plant, which is dependent on the sheep industry in all of its ramifications, and there results an accumulation of many millions of dollars, a value which, from a business stand-point alone, should cause the Government to foster and to protect it from every source of injury.

As the whole growth of the industry is dependent on the health and vigor of the sheep, it follows that whatever tends to produce a better condition or ward off threatening disease from them is for the benefit of all interested in and dependent upon the success of the industry. The parasitic diseases — those produced by the animal parasites of sheep—are, if we may judge from observation and the letters of inquiry directed to this Bureau, the chief source of losses, and if in any way this bulletin may result in promoting a better knowledge of these too little known pests, and in teaching facts which will lead to better care and treatment of the flocks as regards hygienic prevention of diseases, the cost and labor laid out upon the work in its various details will be well expended.

Particular attention has been devoted to illustrating each species of parasite, and, so far as possible, the lesions of the disease produced by

---

* United States Department of Agriculture, report on numbers and values of farm animals, January, February, 1889, pp. 5 and 6.

9

it. In the illustrations of the species certain features which present specific differences have been constantly drawn. Peculiar features of anatomy and development have also been illustrated. The entire development of any species from the unimpregnated ovum to the adult form is not illustrated, but one species may show the developing ova, another the embryo, and still another small and adult forms, and thus the entire development of many of the species can be well understood. Especial attention has been devoted to representing certain organs of economic importance, i. e., those organs which are immediately concerned in injuring the tissues of the sheep. The majority of the drawings were made from nature by Mr. W. S. D. Haines, and the others by Dr. George Marx, both artists connected with this Department. The excellence of their work shows for itself. Where material for original illustration has been unavailable, figures chosen from the leading text-books on the subject under consideration have been copied, and due recognition of the source acknowledged in the description attached. For the accuracy of these drawings the author alone is responsible. He believes that all the anatomical details are accurate, but such is the difficulty of seeing the minuter details that some of the latter are omitted. As their presentation belongs properly to a more specific investigation than this their absence will scarcely be noticed. It has been the constant endeavor of both the artists and the author to make technically perfect drawings, and at the same time present the subject so clearly to the eye that not only a novice may, by the aid of a small magnifying glass, be able to determine the species, but that the scientist may also use the work profitably in subsequent investigations.

The text devoted to each species is intended to contain a general description of the parasite, its life history, the way it causes disease, the disease produced and mode of treatment, both preventive and remedial. Many of the specific descriptions are technical. To the beginner, who can identify the species by careful comparison with the figures, these are unnecessary, but as he advances in their study the meaning of the technical descriptions will become more apparent and useful. In a work of this character such technical specific descriptions are unavoidable. To the scientist they are absolutely necessary. Wherever possible the complete life-history of the parasite is described and illustrated; unfortunately, however, the species whose life-histories are positively known are too few. Although the life-histories of the majority of the worms seem very evident, still the evidence upon which they are based is not deemed entirely conclusive by scientists. So skeptical are the majority of this guild that rigorous proof alone seems to satisfy them, and this is particularly the case when the views set forth in regard to either of the species are at variance with pre-existing opinions.

Rigorous demonstration of the various stages in the life-history of a parasite demands that its eggs or embryos shall be fed to an uninfected host (sheep in this case), and the parasite be found in it subsequently, at a stage of growth corresponding to the time which has elapsed during the experiment. The conditions necessary for raising embryos, for procuring uninfected sheep and for keeping them from outside sources of infection, are many and difficult to fulfill. Up to the present time, with few exceptions, infection has been secured in the experiments only by excluding or regulating certain of the conditions surrounding sheep. These conditions are such that, although the problems in each case have not been absolutely proven, there is much probability that the life-history of most species is well determined. In describing the injury wrought by the parasite and the resulting disease, technical description has been avoided as much as possible, in order that the work may be rendered more valuable for farmers and ranchmen, who have but a limited knowledge of the terms used in medical literature. These descriptions are, on this account, necessarily imperfect from a scientific point of view, but it is hoped none the less efficient for the purpose. A careful study of the various diseases will show that the irritations set up and the lesions resulting therefrom are mainly due to mechanical causes, whatever be the organ attacked. Certain of the diseases, however, seem to be aggravated by nervous or reflex irritation induced by the parasite, while others are hastened by a loss of blood or nutritive material abstracted from the host by the parasites.

The diagnosis of parasitic diseases is always determined by finding the parasite or its eggs. The quickest and surest determination for internal parasites is made at a *post-mortem* examination. For intestinal parasites many authors recommend the examination of the dung. This method has not been verified by experience, but appears to be tedious and difficult, and a method better adapted to experts than layman.

There are certain symptoms from which one may infer that sheep are infected with parasites. A large part or all of the flock is affected and the symptoms shown by the different individuals are similar. The appetite is generally good, but individual members present a poor, stunted, hide-bound, bloodless, big-headed, pot-bellied appearance. Other local symptoms, depending on the organs affected, are present. The most positive characteristic is to find that a number of sheep raised together are affected in the same way. From these general symptoms those depending on climatic changes, and irregularity or insufficiency of food and water, must of course be excluded. The sheep owner who discovers weakness among his lambs should not wait until one of them dies before he endeavors to make a diagnosis, but should undertake to diagnose the disease in the earlier stages by sacrificing one or more of the worst affected, and thus gain time in treating and preventing the extension of the disease. By waiting for the disease to develop he allows the lambs to grow poorer and weaker, and when action is finally undertaken it is upon patients which are, in many cases, already too weak to stand vigorous treatment, and which can in no way profit by preventive measures as they should. The lambs examined can, if the meat is not too poor and watery, be used on the table without harm to the consumer. If the animals are at all feverish, as is the case in the later stages of disease, the carcasses should be thrown away. It is in the beginning of the disease that treatment, both hygienic and medicinal, is needed and produces its best results, and therefore an early diagnosis and determination of the malady is fully as essential as in the more virulent bacterial scourges.

Though the *treatment* advised in a work of this character should be its strongest point, yet it is to be regretted that such is the state of knowledge of the life-history of these parasites and of the practical results of medicines used in combating them under the conditions in which sheep are held on the pastures, that it is felt that this field is yet to be properly entered and worked up from an experimental stand-point. The subject appears, as yet, to be in an empirical stage. Although the best recipes have been compiled and presented, they appear to be old and hackneyed to one who has been enabled to trace the same recipes from book to book. Indeed, some of those presented, which contain inherent virtues, come from countries where sheep-ranching is unheard of, and seem to be sufficient only in the closely settled communities where labor is cheap and where time can be devoted to saving property even though the value is not great. The medical treatment of large flocks should be investigated from a broader stand-point than any yet taken. Our insufficiency of knowledge on these points arises from the small value of single animals and the hesitation of people to seek the aid of skilled veterinarians until they find that they are unable to treat the disease themselves. The great benefit in doctoring animals whose individual worth is but a few dollars lies in the treatment of numbers at a time, and in making an early diagnosis of the disease. Those who have large and valuable flocks should watch their lambs for the earliest symptoms, and then if there is a skilled veterinarian available obtain his services. Oftentimes the family doctor can and will give advice that will materially assist, for his knowledge of other diseases, their symptoms and lesions, and of the use and effects of medicines, make him the most available authority in the absence of the veterinarian.

Upon the hygienic treatment, i. e., upon the care and attention the flock receives, depends in great measure its health and good condition, and the prevention of the parasitic diseases. It is out of the province of this bulletin to discuss the proper housing, food, and drink of sheep, beyond what is required for the prevention of parasitic maladies. The chief necessity as regards buildings and yards is that they should be kept clean. Periodic cleansings of wood-work and floors should be sufficient. Whitewashing and the liberal use of lye water for cleansing wood-work are desirable, and in some diseases, such as scab, absolutely indispensable. In the care of yards an economic management of the manure is to some of prime importance. It would seem that a mixture of this manure with lime in the compost heap, and a frequent cleansing of the yard, would be far better, so far as the sheep are concerned, than to allow it to accumulate. The lime would not only serve to kill the eggs of parasites in the manure, but would add fertilizing material to it. Manure so treated would be a better fertilizer, and would also be less apt to infect sheep when spread upon the fields. The compost heap should never be where the liquor from it can be washed by the rains into water which the sheep drink. As the manure from these yards may prove the source of infection, sheep should never be pastured on fields recently enriched with it, unless there is absolute certainty that the previous treatment of the manure has destroyed all the embryos of the parasites. As frequently urged in the text, every means should be taken to supply sheep with pure water. Although experiments show that sheep have other means of getting parasites than from the water they drink, yet this is at times a very fertile scource of infection. The use of drinking-troughs into which water runs or is pumped, and rapidly running water, seem best suited to meet the requirements.

The grain food should be fed from cleanly swept troughs or floors. Hay should be put in racks, as feeding from the ground is not only wasteful but tends to infect with parasites. Salt should be supplied in boxes placed where sheep can have ready access to it. The mixture of a small proportion of finely powdered sulphate of iron with the salt is allowable at times.

Pastures, which are ordinarily uncared for further than to provide fences for securely confining the sheep, need careful supervision. Wet swails, bogs and swamps should either be fenced out or drained. Pastures which are overstocked, and in which a flock of sheep is kept continuously, are the most fertile sources of infection. Not only do the sheep become more frequently infected where they are compelled to eat the grass close to the ground, but the chances of their being compelled to graze on an infected area are largely increased by keeping them ranging over the same ground of limited area week after week. Old sheep stand such treatment much better than young ones. For the latter, those fields which have not been pastured on by older sheep are better. The practice of feeding the sheep over fields from which the crops are removed is a good one, not only for the sheep but for the fields. These remarks, of course, apply more strictly to fenced farms and not to unfenced sheep ranges, but even on these certain portions of the range can be reserved for the lambs. The practice of allowing lambs, after they are old enough to wean, to feed after older sheep is also a source of infection.

The relation of the dog to sheep husbandry is too important to be overlooked. Were it not that the definition of parasite excludes such animals as can be considered beasts of prey, the dog would be placed at the head of the list of parasites as being the most destructive. Though this be unmistakably apparent to a large majority of sheep-owners, there are many who believe that the dog is man's most faithful friend and that he is of great use even on a sheep farm. It is unfortunate for the dog that the mass of testimony on this subject is against him. It is not from the stand-point of *the dog as a beast of prey*, however, that this work is written, but it is from the more technical stand-point of *the dog as a carrier of parasites dangerous to sheep and man*. In the list of parasites of sheep there are at least four which are common to the dog and sheep, viz: *Tænia marginata*, Batsch; *T. cœnurus*, Küch.; *T. echinococcus*, v. Siebold, and *Linguatula tænioides*, Rud. The last is rare, and in justice to the dog should not be used against him, although it may subsequently afford as damaging evidence as the other species. By referring to the descriptions of the other three species it will be found that dogs harbor in their intestines the adults of these species, and that they scatter the eggs of the parasites broadcast for the infection of sheep. Thus each dog, harboring one or more, is a constant menace to the health and lives of the flocks in the neighborhood. Nor is this all, for man himself can be infected by at least two of these species—*Tænia echinococcus* and *T. marginata*—in their cystic stage. The former of these species produces a disease of slow development, but one which is nearly always fatal in results. To prevent these diseases the precautions prescribed in the text must be closely adhered to. A plan which would remove much of the loss caused by dogs by doing away with them entirely is scarcely practical in this country, where the majority of these animals are owned by persons who have no direct interest in sheep. The hunting and the sheep dog are most to be feared, unless we except some of the fiercer watch-dogs which are kept at slaughter-houses and fed on waste bits. The day of the usefulness of hunting dogs is quite past, and their retention by sheep-men at least should be abandoned. The watch-dogs are nearly always chained and in places not accessible to sheep.

In the range country the coyotes and prairie wolves still menace the flocks by killing individuals for food, and by harboring the adults of *Tænia marginata* and *T. cœnurus*, the eggs of which they also scatter. In most sections, especially where a bounty is offered for their scalps, the trap is remorsely exterminating them. Laws which would subject the mongrel curs to the same treatment would result in a great gain to the farming community and to their respective owners, if they be owned by any one.

There are reports that the sheep can be infected by parasites from some of the many wild animals that still haunt the land where they were formerly so abundant. The examination of these little quadrupeds to ascertain the parasites they contain has not been as extensive as it should be for a broad generalization, but so far as it has extended it is safe to say that sheep are not infected from either rabbits, skunks, squirrels, woodchucks, gophers, prairie dogs, or foxes. Foxes may harbor some of the adult *Tænia* whose cystic stages infest sheep, but unless they can obtain the young forms of the parasites by eating the viscera of sheep they would be very unlikely to be infected with adults. It is also reported that rabbits harbor the cystic form of *Taenia marginata*; but this statement has not been corroborated, as the rabits examined contained the cystic form of *T. serrata*. If rabbits should be proven to harbor the cysts of *T. marginata* then the danger would arise, not from them, but from hunting-dogs which eat the rabbits and the cysts they contain, and harbor the adult parasites that alone are the source of danger. The possibility of infection by parasites from deer is too small to be considered as an economic question, owing to the great scarcity of these animals. The antelope (*Antilocapra americana*, Ord.) may be a bearer of many of the same species of parasites as the sheep, but they also are becoming too few to be considered as a source of danger. In short, it is futile for the flockmaster to consider these sources at all while his own flocks are infecting his fields, and his dogs are constantly assisting them. Could these sources of infection be controlled, there would be no need for him to regard the wild animals as his enemies.

In purchasing sheep particular attention should be paid to the general appearance and past history of the flock from which the purchase is made in order to avoid parasitic diseases. Before adding recent purchases to flocks they should all be thoroughly dipped to kill external parasites. If they are coughing ever so slightly, the cause of the cough should be investigated to determine the presence or absence of lung worms. If some are hidebound or weak after allowing for the character of the season and the condition of pasturage the possible presence of intestinal parasites should be next considered. It is not very probable that there are any farms free from all parasites, but there are many that are free from a considerable portion of the species which are properly parasitic on sheep. Purchasing here and there in making up a flock brings all sorts of parasites together, thus infecting a farm to such a degree that it is difficult to get rid of them.

The medical treatment must, of course, be specially adapted to the disease. The treatment of external parasites is effective, and well repays all efforts. The treatment of internal parasites may be divided in general into treatment for lung worms, for intestinal worms, and for liver worms. The last is by far the most unproductive of good results. Parasites situated elsewhere in the sheep do not readily yield to medical treatment.

Scab is the only parasitic disease that has been thought worthy of legislation. There are others that demand as serious consideration, but their importance has not yet been fully presented to the public. Little attention has been given to police interference in the management of these diseases. No doubt such interference might be profitably pushed further than it is in this country, especially with regard to scab. Not only should the highways be guarded against the movement over them of scabby sheep, but a competent innspector should be appointed by the State to supervise every sheep dipping, to compel the dipping of every scabby flock, and to attend to the renovation and disinfection of the sheep-yards and walks. Every band of scabby sheep is a constant menace to the health of others. In this country there seems to be no sheep disease produced by animal parasites which renders the flesh harmful to man, further than that some of the flesh may be less nutritious. Until the sale of meat of all kinds is guarded by more stringent regulations there does not seem to be any reason for urging police restrictions on the sale of meats of the inferior quality which some of these diseased lambs undoubtedly furnish.

There are described in this volume twenty-six species of animal parasites of sheep, as follows:

1. Œstrus ovis.
2. Melophagus orinus.
3. Trichodectes sphærocephalus.
4. Trichodectes climax.
5. Trichodectes limbatus.
6. Psoroptes communis.
7. Linguatula tænioides.
8. Tænia fimbriata.
9. Tænia expansa.
10. Tænia marginata.
11. Tænia tenella.
12. Tænia cœnurus.
13. Tænia echinococcus.
14. Distoma hepaticum.
15. Amphistoma conicum.
16. Distoma lanceolatum.
17. Strongylus contortus.
18. Strongylus filicollis.
19. Strongylus rentricosus.
20. Ascaris lumbricoides.
21. Dochmius cernuus.
22. Sclerostoma hypostomum.
23. Œsophagostoma Columbianum.
24. Trichocephalus affinis.
25. Strongylus ovis-pulmonalis.
26. Strongylus filaria.

Of the species described three genera—*Melophagus, Trichodectes,* and *Psoroptes,* embracing five species, *M. orinus, T. sphærocephalus, T. climax, T. limbatus,* and *P. communis*—are external parasites.

The species which there is reason to think do not occur in this country are *Tænia tenella* and *Amphistoma conicum.* The former is considered by continental authorities as a synonym of *T. solium* or *T. marginata.* The writer has not found *Linguatula tænioides, Tænia cœnurus, T. echinococcus, Distoma hepaticum,* or *D. lanceolatum* in sheep, nor learned from authentic sources of any of these occurring here except *D. hepaticum.* The other species may eventually be found, but they will probably be rare. One other species, *Ascaris lumbricoides,* seems to be a rare one in sheep. The remaining species are all more or less abundant. *Tænia fimbriata* and *Œsophagostoma Columbianum* seem to be exclusively American species. The others are common to all parts of the world where there are sheep. Von Linstow (*Compendium der Helminthology,* 1878), catalogues nineteen species of internal parasites which infest European sheep. One of these, *Monodontus Wedlii,* Molin, is a synonym of *Dochmius cernuus* Creplin; another, *Nematoideum ovis,* Diesing, is a lung-worm insufficiently described. Still another, *Trichosoma papillosum* Wedl., is a synonym of *Strongylus filicollis,* Rud. The remaining sixteen species are described in this volume. There are two lung parasites of sheep in the Old World that have not been found here, viz: *Strongylus rufescens*, Leuckart, and *S. minutissimus*, Megnin. The former is said to occur in Germany and France, but is so meagerly described that it probably is not a distinct species, but a synonym of *Strongylus* (*Pseudalius*) *ovis-pulmonalis*, Diesing. The latter occurs in Algeria, is well described and figured, and seems to be a well established species. *Strongylus ventricosus*, also a European species, has not, to my knowledge, been described as a parasite of sheep heretofore.

A further comparison of the above list with those of parasites of sheep in other countries is re-assuring, because, first, native sheep have now nearly all the parasites that they can acquire in this country; second, that although nearly all the European species have been imported, *Distoma hepaticum*, L., the liver-fluke, *Tænia echinococcus*, v. Siebold, and *T. cœnurus*, Küch., are either very rare or else do not exist in this country. These three parasites have been the cause of great loss among sheep in other parts of the world.

The comparatively long list of parasites furnished will seem to the European to indicate that sheep in this country are more infected than those in Europe; but in this connection it should be remembered that much time has been spent in hunting for several of these species, and some of them are rare, inconspicuous, and do little damage.

The following is a list of our most destructive ovine parasites: *Œstrus oris*, L.; *Psoroptes communis*, Fürst.; *Tænia fimbriata*, Diesing; *T. expansa*, Rud.; *Strongylus contortus*, Rud.; *Dochmius cernuus*, Creplin; *Œsophagostoma Columbianum*, Curtice; *Strongylus ovis-pulmonalis*, Diesing; and *S. filaria*, Rud. There are nine species in all, a list which compares favorably with that of the ovine parasites of any other country; for all but two species, *T. fimbriata* and *O. Columbianum*, are common to all countries, and these two are more than offset by the prevalence of more destructive parasites in the Old World.

On the whole, the flockmasters of the United States may be congratulated on the good condition of their flocks and their comparative freedom from both external and internal parasites.

PARASITISM.

Definition.—The animal parasites of sheep are those which live in or upon their living bodies and obtain nourishment from them. The term "animal parasites" is used in order to distinctly separate this group from the vegetable parasites which attack the living organs of sheep. Both animal and vegetable parasites prey upon the flocks and cause disease, but such are the differences between them, their effects and the methods of investigating them, that an investigation of either forms a large field of research.

Parasites as defined above include a large number of animals so different from one another that parasitism is the only common character which groups them together. Though this distinctive feature is sufficient for the present purpose, it is a very variable character, for the degree of parasitism manifested by each of the species varies through all the scale possible from the transient momentary parasites to the permanent.

The animal parasites of sheep are all embraced within three great branches of the animal kingdom : The *Protozoa*, *Vermes*, and *Arthropoda*. None of the first branch, which includes the *Coccidia* and *Balbiana gigantea*, Raillet, are described in this volume. Examples of the second, which includes all the worms, and of the last, which includes the insects, mites, and linguatula, are abundant.

The worms live, as a rule, in the open cavities of the body—in air spaces of the lungs, the ducts of the liver, and the lumen of the intestine. The exceptions to this rule arise from those immature forms which penetrate into the substance and closed cavities of the bodies, *e. g.*, the bladder stages of the tape-worms and the young embryos of *Œsophagostoma*. The worms are called *internal* parasites. It is easily understood, however, that being held in the cavities of the body which have communication with the exterior, they are really external to the body tissues, and only those embryonic forms which penetrate into the tissues of the infested animal or host are true internal parasites.

As a rule the insects live on the surface of the body. They are called *external* parasites. The exceptions are the larva of *Œstrus ovis*, which lives in the nasal cavities, and *Linguatula*, whose young stage infests various organs of sheep. *Œstrus* is usually classed with the external parasites, and *Linguatula* is in sheep truly internal. As has been stated, those parasites which in their young stages penetrate the tissues of sheep are alone true internal parasites. Even these spend their adult stages in the open cavities of some other host and then become true external parasites, so that no one of these parasites is, strictly speaking, an internal one throughout its life. That every parasite should be an external one in its adult stage is a necessary condition of its existence and of the perpetuation of its species, for it is only in the open cavities that they can obtain sufficient air and food, and can mate. From these cavities, too, the eggs and young can escape for the infection of other sheep. The facility offered for mating and distribution is the most important reason. In order to avoid confusion of terms those parasites infesting the surface of the body will be called, in conformity with custom, *external* parasites; the others, which inhabit the tissues of the body and its cavities, *internal*.

Though the animal organisms that infest the living bodies of sheep be small, they are endowed with all the vital functions of life. All can move, feed, feel, and reproduce. None of the worms can see or hear. The insects are more highly specialized than the worms. All of them have in the past become so adapted to their surroundings that they can live in no other, and while sheep thrive better if not infested by parasites the latter can not live without sheep. The only exceptions are those species which are also parasitic on other animals, as goats and cattle. The modifications of organs which have arisen out of the needs of parasitism are too many to give in detail. The great central fact of their lives is that all the parasites have arisen from their kind, and under favorable circumstances will reproduce their species, and that they are to be treated as the originators of disease and not as the products of disease.

The methods by which sheep become infested differ with the species. The external parasites are usually transmitted by actual contact of sheep against sheep. The parasites may, however, be dislodged from their former host and afterwards make their way to another sheep. The first is known as mediate, and the second as immediate contact. The diseases produced by the external parasites are true contagious diseases, and should be regarded as such fully as much as any of the more actively virulent maladies. The transmission with this class of parasites is usually an active one; they may, however, be borne from one sheep to another by people, cattle, goats, or by locks of wool, when the transmission would be passive.

Œstrus ovis, which seems to bridge the gap between the external and adult internal parasites, differs from these groups in being able to actively infest its host with its young, without an actual contact or intermediate bearer. Lice, louse-flies, and scab insects may do this in a less degree, but not to that possessed by the *Œstrus*. The *Œstrus* larvae are never transmitted by contact; they must mature, fall to the ground, metamorphose, and emerge as adults before the females can infect sheep. The internal parasites are passively conveyed into sheep along with the food and drink consumed, and never actively enter into the transmission. They may be conveyed either as eggs or very young embryos. *Œstrus* forms the single exception.

The terms "*contagious*" and "*infectious*" can be applied to these parasites. The former is applicable to those parasites which usually transmit themselves to other hosts, the latter to those which are transmitted to their hosts along with food and drink. The young of *Œstrus* have no agency in their transmission, and hence infect sheep.

Parasites are frequently said to *invade* the hosts which harbor them. This is only true of those species which actively undertake migration, as scab insects and sheep ticks. A few species invade the organs of their hosts after the latter have been infected, thus : The larvæ of *Œstrus* crawl from the margins of the nostrils to the sinuses of the head ; the lung worms migrate into the lungs; the young embryos of *Tænia marginata* tunnel the liver; *T. cœnurus* tunnels the brain; *Œsophagostoma* penetrates into the intestinal walls. Those internal parasites which undertake active migration in the bodies of their hosts seem to form a minor class in the parasitic world, those which lodge in the intestine and ducts emptying into it forming the majority.

The ability to *select* their final lodgment belongs to each species, and is the one character on which their own life and that of the species depends. This is self-evident in the case of external parasites. After hosts are once infected by the internal parasites and the young embryos are endowed with activity, they either select their proper place while being carried along by intestinal fluids, or force their way to it through all opposing tissues and against all counter currents of fluids. Those embryos which fail to reach these places finally die for want of the necessary conditions of life. The very ability that is so absolutely necessary to enable certain of the parasites to reach their chosen organ often proves the means of their premature death. *Tænia marginata* cysts invading the liver become lost in the mass of this organ and perish. Multitudes of these parasites injure the capsule of the liver and cause the sheep invaded to die long before they have matured sufficiently to pass into dogs. The embryos of *Œsophagostoma* often wander into the mesenteries, the retro-peritoneal glands and liver, and perish.

Parasites escape from their ovine hosts either actively, e. g., the young and adults of the louse-flies, lice, mites, and the larvæ of *Œstrus*, or passively as eggs or young embryos, the young embryos of the *Strongylus filaria* and *Tænia expansa*, the completely segmented eggs of the *Strongylus contortus*, and as eggs incompletely segmented. In the latter case they are rejected with the excreta of the lungs or intestines. A very few (the cystic tape-worms) escape only after the death of their host by the intervention of some carnivorous animal which swallows them with its food and liberates them from their imprisonment by the processes of digestion. The death of the host is usually caused by the carnivora in search of their food. The continuance of the parasites' life into the adult stages depends, therefore, on the destruction of their host. This fact is contrary to the usual rule of parasitism, which demands that the host continues to live in order that the parasite may live and reproduce its species.

The length of time and the stage of development at which parasites *infest* their host varies considerably. Lambs have no parasites at birth. Within a month or two after, they become infested by a few individuals of certain species of round worms, and by external parasites. From this time on they may harbor any of the species to which they become exposed. It will be noticed that the commencement of infection begins when the lambs first nibble grass. The louse-flies, lice, and scab insects infest the fleeces and skin from generation to generation. Unless it should subsequently be proven that the hair-lungworm (*Strongylus ovis-pulmonalis*), and the stomach round worm (*Strongylus contortus*), may also perpetually infest sheep, they harbor no other species throughout their entire life cycles. *Œstrus ovis* is parasitical only in its larval stage, and consumes months in developing. Because it can not take nourishment when adult, it is believed to pass a very ephemeral adult stage. The broad tape-worm develops rapidly and disappears, its six-hooked embryo apparently spending long seasons of suspended life functions on the ground. The fimbriated tape-worm develops more slowly, consuming the greater portion of the year; its embryos may exist on the ground for indefinite periods. The cystic tape-worms pass indefinite periods as cysts in sheep, depending on their resistance to the vital forces of the organs infested and upon the date of their liberation from imprisonment. The life cycle of the liver flukes seems to be completed in a few months. The majority of the round worms seem capable of withstanding the elements while scattered over the pastures for indefinite periods, either as ova or partially and completely developed embryos. Their cycle of life in sheep is of variable periods, depending on the species. (*Esophagostoma*, some of whose embryos invade the intestinal wall, offers a retarded development lasting through months. Other species develop more rapidly. The exact cycle for each species has not been determined, but most of them become adult in less than six months, some in less time than three.

The *seasonal appearance* of each species depends on its life cycle, the average temperature and the humidity of the season, and the age of the lambs. Spring and fall seem to offer the most outbreaks of disease produced by parasites. Summer and winter also have their special parasitical diseases. Sheep-ticks, lice, and scab are more prevalent in winter when the sheep are closely herded in yards or barns, and when they are covered with heavy fleeces. The gad-fly occurs most in June and July, but in milder climates it evidently flies the greater part of the year. The disease it develops is more prevalent in older sheep, yearlings being the youngest that show distinct signs. The broad tape-worm infests young lambs early and causes their disease in a very few months. March lambs harbor adult worms in May and June, and May lambs in August. The fimbriated tape-worm also infests lambs early, but does not produce its worst effects until late fall and winter. The liver flukes generally appear first in summer and fall. The round worms appear in young sheep of three months and upwards. The majority of those that produce disease develop it as they grow adult. The thread lung-worms (*Strongylus filaria*) infest lambs, and epidemics due to them usually occur from spring to fall. The hair lung-worm, on the other hand, develops slowly, and while their presence can be detected in the lungs of young lambs it is the lungs of old sheep which show the greatest amount of changes due to their invasion. As a rule warm, moist seasons are most favorable to their development. The climate of the United States so varies from North to South and East to West that no exact seasonal appearances of the various species can be given. Most of the species seem to be present in sheep in greater or less numbers the year round. The most important factors in the time of outbreak of different diseases seem to be the age of the hosts and the cycle of life of the parasite.

The destructiveness of each species is dependent on the numbers of the invading parasites, the organ invaded, the method by which they produce disease, and the age of the host. As a rule, most parasites produce disease by their numbers, each causing its infinitesimal amount of annoyance. The sheep-grub, the broad tape-worms, *Dochmius cernuus*, and *Tænia cœnurus*, are notable exceptions to the rule. But few individuals of each of these species are found invading the organs of sheep. Their destructiveness depends on the character of the annoyance produced and on the organ invaded. A few of nearly all species may infest sheep, and seemingly cause no loss, but when any of the factors favoring the development of either of the species appears they increase innumerably and destroy the lives of their hosts.

Parasites effect injury to the health of sheep in many ways, some of which are very evident to all, while others are indefinable and illusive. The injuries effected by sheep-scab, by the hair lung-worms, and by *Œsophagostoma*, are easily discovered; but the injuries produced by the tape-worms and various species of round worms can not always be detected in the intestinal walls. In these the method of determining disease is not by testing the various organs by the microscope or chemical analysis, but by comparing the patient with a healthy animal of the same age. in action, appearance, weight, etc., and by comparing the organs of the patient with an actual or assumed standard. In these ways the effects of the subtle diseases produced by the parasites are learned and just allowance made for their importance.

Most parasites mechanically injure tissues. They may either force their way through soft tissues as the cysts of *Tænia marginata*, break through the structure of lung tissue as *Strongylus ovis-pulmonalis*, enter into the intestinal walls as *Œsophagostoma*, or lacerate the tender mucous membrane as *Dochmius cernuus*. The adult tape-worms seem to irritate the intestines, derange their functions, and cause nervous disturbances. *Dochmius cernuus* and perhaps *Strongylus contortus* seem to abstract blood. Other round worms may live on the intestinal contents.

The destruction caused by the different diseases produced by parasites is varied. Scab, lung-worms, flukes, and tape-worms often destroy entire flocks. The ravages of other species are less patent. There is no doubt, however, that each causes more trouble than has yet been assigned to it. The least destructive of the species common to sheep in this country seems to be *Trichodectes sphærocephalus*, *Strongylus filicollis*, *S. ventricosus*, and *Trichocephalus affinis*.

No rigid system has been adhered to in the succeeding pages. Nevertheless, a certain plan of arrangement and treatment will be noticed. The highest insects in the zoologic classification precede. These are followed by the *Acarina* or scab mites. The Helminths, embracing the *Platoda* or flat worms, and the *Nematoda* or round worms, follow in turn. A systematic classification of the worms among each other is not closely adhered to. It may be noticed further that such an arrangement permits of a second subsidiary treatment of the subject by grouping the parasites together as regards the organs they infest. The variations arising are due in part to the antagonism between the two arrangements adopted. The above plan was adopted more for convenience than for its scientific accuracy from any single point of view.

Concluding, the author desires to give due recognition to all sources, from which he has drawn in preparing this work. The chief sources of information have been the admirable works of Leuckart, *Die Menschlichen Parasiten*, 1868; Raillet, *Elements de Zoologie*, 1885; Neumann, *Traité des Maladies Parasitaires*, 1888; and A. E. Verrill, *Parasites of Domestic Animals*, Rep. Conn. Board Agriculture, 1870–'71. An endeavor has been made to mention every other source in the text.

MEASUREMENTS.

The following tables will assist the reader in reducing the measurements given in one denomination to those in another.

Metric system in medicine.

	Grams.
1 grain or 1 minim equals	0. 06
15 grains or 15 minims equal	1. 00
1 dram or 1 fluid dram equals	4. 00
1 ounce or 1 fluid ounce equals	32. 00

The cubic centimeter may be considered identical with the gram for water or aqueous solutions.

Metric system in measurement.

	Inch.		
1 meter equals	39. 37		
1 centimeter equals	. 3937	or	$\frac{2}{5}$
1 millimeter equals	. 03937	or	$\frac{1}{25}$
One-hundredth millimeter equals	. 00039	or	$\frac{1}{2500}$
One-thousandth millimeter equals	. 000039	or	$\frac{1}{25000}$
25 millimeters equal	1 nearly.		

OTHER MEASUREMENTS.

For approximate measurements a liter (2.113 pints) may be considered equivalent to a quart; a kilogram to $2\frac{1}{5}$ pounds avoirdupois.

When graduated measures of weight or volume are not at hand, the flock-master may use some of the common household utensils. Pint and quart bottles, so called, should be tested, as they vary in size. The pint contains 16 ounces, or about a pound in weight. The smaller bottles are known as 2, 4, 6, 8, and 12 ounce bottles. Vials are quite common, those made to contain 1, 2, and 4 drams being most abundant. A set of bottles can always be obtained at the drug-store and the size marked on them. It is far better, however, to buy a set of graduates and other measures, for they are of daily use. Common tumblers contain from 8 to 10 fluid ounces; tea-cups, about 5 fluid onces; wine-glasses, about 2 fluid onuces; tablespoons, half a fluid ounce; dessert spoons, 2 fluid drams, and teaspoons, 1 fluid dram. .

In spite of the fact that the use of the metric system has been legalized in the United States, so deeply is the old English system engrafted upon the customs of our people that there will necessarily be a confusion of weights and measures until the use of the former is made compulsory. This system is so easy to learn and so easy in practical application that it will certainly supersede the other in use, as it has in the continental countries of Europe.

THE SHEEP GADFLY—GRUB IN THE HEAD—NASAL CATARRH.

ŒSTRUS OVIS, Linn.

Plates I, II, and III.

The popular names of this well-known parasite of sheep convey to the reader its epitomized life history, long known to veterinarians and farmers. Though the life history is a comparatively simple one, there are many of its details which are not only unfamiliar to the average shepherd, but some of which are unknown even to those who have made a special study of these pests.

The "Sheep Gadfly," the parent of the "Grub in the Head," is, when flying, so small and so quick in its actions that it is very difficult to see, and still more difficult to catch. The greater proportion of specimens in collections have been raised from the grubs, *i. e.*, the adult grubs are collected and placed in a net-covered box, the bottom of which is covered by a couple of inches of damp sand. In a few minutes they bury themselves in the sand, and in from three to four weeks they reappear as flies.

C. V. Riley (*Insects Missouri*, First Annual Report, 1868, p. 161) describes the fly (Plate I, Fig. 11) as follows:

In this stage it looks something like an overgrown house-fly. The ground color of the upper part of the head and thorax is dull yellow, but they are so covered with little round, elevated black spots and atoms (scarcely distinguishable without the aid of a magnifier) that they have a brown appearance. The abdomen consists of five rings, is velvety and variegated with dark brown and straw color. On the under side it is of the same color, but not variegated in the same way, there being a dark spot in the middle of each ring. The feet are brown. The under side of the head is puffed out and white. The antennæ are extremely small and spring from two lobes which are sunk into a cavity at the anterior and under part of the head. The eyes are purplish brown, and three small eyelets are distinctly visible on the top of the head. It has no mouth, and can not therefore take any nourishment. The wings are transparent and extend beyond the body, and the winglets, which are quite large and white, entirely cover the poisers. Its only instinct seems to be the continuation of its kind. It is quite lazy, and except when attempting to deposit its young its wings are seldom used.

The male is about as large as the female, but may be known by its relatively narrower forehead or space between the eyes. Catching a male at large would be a chance operation, for though they fly to mate with the females around the sheep-yards and pastures, they never make their presence known by disturbing sheep.

25

Brauer (*Monographie der Œstriden*, p. 15) records the size of the male and female at from 10 to 12^{mm}, or about two fifths of an inch; the width between the eyes of the male at 1^{mm} or one-twenty-fifth of an inch, and of the female at 2.5^{mm} or one-tenth of an inch. The length of the wing is 9^{mm}, nearly two-fifths of an inch.

Distribution.—The species occurs all over the world wherever there are sheep. It is now too late to learn if it was indigenous in this country, but we may believe that it was introduced with the earliest flocks imported, whether in Mexico or on the Atlantic coast. Brauer, in 1863, stated that it had then lately been introduced into Chili, South America.

Life history and description of larva.—The interest of the flock-master in this species begins when the fly buzzes around the noses of the sheep and deposits its young just within the opening of the nostril. Many of the older writers on this subject supposed that the fly deposited eggs, but Brauer (*o. c.*, p. 151), in agreement with Joly and Dufour, pointed out that the genus was one which deposited their young alive. According to Riley (*o. c.*, p. 164) Samuel P. Boardman, of Lincoln, Ill., mentioned the two following independent discoveries: John Brown, "Old Ossawattomie," stated in the Ohio Farmer about 1851 that he saw the fly drop the perfectly formed and living grub in the nostrils of sheep. About 1861 Dan Kelly, of Du Page County, Ill., made the same discovery, and records the fact in the Prairie Farmer, October 14, 1865. Boardman, in 1867, received a letter from Mark Cockrill, of Tennessee, who wrote of having made the discovery years previous. Riley claims (*o. c.*, p. 165) that he obtained living maggots from a fly in 1866.

The young larva, having been deposited within the rim of the sheep's nose, soon attaches itself by means of the hooks (Plate I, Fig. 6), and begins to make its way upward into the nostrils. The smallest specimens collected by the writer are shown in Fig. 7*e*, natural size, and are not much larger than when first deposited; for the difference between their size and that of eggs deposited by flies even smaller than *Œstrus ovis* is inconsiderable. These small larvæ are in their first stage of growth. They are little, white, elongated bodies, less than 2^{mm} long, *i. e.*, about one-twelfth of an inch (Figs. 1 and 2). But little of their structure can be seen except with a lens. They already show the division of the body into eleven segments, two well-defined hooks (Fig. 3*a, a*), and two minute terminal breathing pores (Fig. 4*a, a*). The ventral surfaces (Fig. 2) show the little spines which later on are to become strong thorns; some of the spines on the sides are relatively bristle-like and longer in proportion than they are later on. In their second stage of growth (Fig. 7*c, c*) all the characters are well defined. The skin is white and so translucent that the digestive organs, the respiratory apparatus, and the fine filiform nerves and their ganglia can be readily made out; the spines on the abdomen, the hooks, and the stigmata are all more pronounced. In the third stage—that of the mature larva ready to undergo its change into a fly (Fig. 7a, a, a, and b)—the characters outlined in its first stages have been perfected. It is this stage that is commonly seen by those who split open the head of an affected sheep. The mature grub averages over 20^{mm} in length and 7^{mm} in width, or about three-fourths of an inch long by one-third of an inch wide. Its width and length when measured depend much upon its state of contraction. Its back is very convex, its abdomen is slightly curved but generally flat, its outline is a very elongated oval, with an acute head and obtuse posterior end. Half-grown specimens are more pointed at the ends and decidedly flatter on the abdomen. From the young to the mature larval state there is a decided change of color. At first they are white and semi-transparent. They quickly grow whiter and soon after take on a tinge of yellow, which, as they mature, grows darker and darker. Then, too, on the back of every segment, except the first and last, a dark narrow band appears which eventually changes from a brown black to a dead black. These bands are rather narrower in front, increasing in width backward. On the side of each segment below these bands there is, in mature specimens, a row of dark dots. The spines show the same changes, at first white, then tipped with brown, and finally changed to black. The spines occur only on the abdomen and the closely adjacent edge. They all point backward and assist the grub in moving around.

When the larva is mature it escapes from the nostrils, falls to the ground, bores into it for an inch or two, and, according to Riley (o. c., p. 162), contracts during the next forty-eight hours to half its former size, becomes smooth and hard and of a black color, tapering as in the larva towards the head (Fig. 14). It remains in this state three or four weeks, or according to some authorities from fifty to sixty days, depending on the weather. When the fly has matured within the case it pushes off a little round cap, ascends from the ground through the hole left by the larva in its descent, and emerges into day to complete the cycle of its existence.

The office of the fly seems to be merely to reproduce its kind. On account of its rudimentary mouth it is unable even to eat. After emerging the fly crawls upon some neighboring grass or twig and rests there until its wings and body have hardened. During the first part of its life, according to Brauer (o. c., p. 149), it is very sluggish, sitting around in the cracks and crevices of the walls of sheep stalls, and is so dull that it must be dragged out. When placed on the hand it seldom flies off. This dullness vanishes as soon as the fly has reached perfection and the right temperature of the air comes. It is then off with a whirr, first vertically in the air, and then in the direction of the flocks.

The effects it produces on sheep and how and where the injury is inflicted may now be stated.

Bracy Clark, an English veterinarian, describes the effect of the attack of the fly in Transactions Linnean Society, 1797, vol. III, p. 315, as follows :

Early symptoms.—The moment the fly touches this part (the nose) of the sheep, they shake their heads and strike the ground violently with their fore-feet. At the same time holding their noses close to the earth they run away, looking about them on every side, to see if the fly pursues; they also smell to the grass as they go, lest one should be lying in wait for them. If they observe one they gallop back, or take some other direction. As they can not, like the horses, take refuge in the water, they have recourse to a rut, dry dusty road or gravel pits where they crowd together during the heat of the day with their noses held close to the ground, which renders it difficult for the fly conveniently to get at the nostril.

This description of the action of the sheep when attacked by the fly is correct in all but one or two minor points. The sheep's actions when running indicate that they are taking every means to shake off and dodge a single rather than a number of pursuers. If this quotation had also described the sheep huddled under buildings, along fences, under rock ledges and shade trees, holding their noses close under their fellows, it would have completed a picture familiar to every farmer or flock-master.

The fly only works during the heat of the day, while early in the morning and late in the evening the sheep seem to enjoy feeding in freedom from its annoyance.

Pathology.—The young larva deposited in the nostril of the sheep immediately begins its migrations upward into the dark passages of the nose. It progresses by means of its hooks and spines. By firmly fixing the hooks into the mucous membrane it is enabled to draw up the rear part of its body after it, and by pushing upon the spines of the abdomen it holds itself in place while it thrusts out its head for a new hold. The only appearance of limbs that the larva has is the two rows of prominences along each side of the abdomen, as shown in Fig 13. In this method of progression lies one cause of irritation to the sheep, viz., the hooks sink into the mucous membrane and not only irritate it, but cause minute points of hemorrhage which are afterwards indicated by very minute black dots scattered over the surface of the nares or internal nose. Whoever has felt the tickling and itching of a fly at the entrance to, or a foreign body in the nose, can imagine some of the sensations which induce the sheep to make such attempts to escape its foes.

As the larva grows in size it finds its way farther into the recesses of the nose, and by following the grooved passage (see Plate III, Figs. 1 and 2) penetrates into the furthest chambers. Fig. 1, *n, n,* shows the young larvæ wandering over the turbinated bones and in the main passage. One of these larvæ, advanced in size, is shown at *i* in Figs. 1 and 2 following the direction of the channel marked by the straw *o, o,* which emerges into the frontal sinus of the head, *k.* The larvæ may also wander among the windings of the superior turbinated bones *g,* and finally growing to such a size that they can not escape, become entrapped there. The same may happen when they wander through the small orifice, near but below that leading into the frontal sinus, which leads into the superior maxillary sinus. (See Plate II, *a*, *b*.) A bit of straw has been inserted into the orifice to show the place of opening. From these places the larvæ never emerge, but after maturing undergo calcareous degeneration. Those that arrive in the frontal sinuses seem to thrive, and at the proper time are able to retreat through the orifice they first traversed and are finally sneezed to the ground. This history and the figures illustrating it, which have been drawn from nature, should satisfy sheepmen who have thought that because the grubs were in the head they must be in the brain. A glance at Plate III, with its three larvæ in the sinus (there were originally seven in the head which the artist figured), will show that there is a bony partition, *a*, *a*, *a*, between the brain and the larvæ. This is also the case in regard to those small larvæ which wander among the intricate windings of the upper turbinated bone, *g*. The larvæ of the *Œstrus*, or the grubs, never do and never can penetrate into the brain.

If one may judge from the black dots indicating a previous hemorrhage scattered over the mucous membrane, the irritation set up by the wandering embryos is very considerable. In the sinuses of heads which contain older embryos other changes are to be noticed. They are filled with catarrhal matter which has been produced by the irritation of the larvæ, and the mucous membrane is greatly thickened. These changes may also be observed over the turbinated bones, *g*, the greatest changes occurring in the superior, the one next the brain. The membrane which covers the latter is the one in which the nerves of the sense of smell are distributed, and a thickening of the membrane must greatly interfere with this sense. This is no small matter, for it is mainly by this sense that the sheep separates its food from other herbage.

In addition to the catarrhal product and thickening of the membrane, it has been noticed that the membrane near the base of the turbinated bone—near *m*—is sometimes very dark colored. At this point the bone is exceedingly thin and pierced by a number of holes through which the olfactory nerves, or nerves of the sense of smell, pass. It is not unusual to find the membranes of the brain in the immediate vicinity blackened by minute dots, indicating a previous inflammation at this point. This affection of the membranes is probably caused by an extension of the inflammatory process from the nasal cavity.

When the larvæ become entrapped in the maxillary sinus they excite the same catarrhal secretions and thickening of the membranes as elsewhere, which finally fill the cavity, the outlet of which is at the top. All this irritation is due to the insertion of the claws and scratchings of the abdominal spines. The larvæ live on the material in which they move. They seem to obtain plenty of air even in the most crowded recesses. They continually cover and uncover the breathing pores (Plate I, Fig. 10*a*), and in so doing keep them cleaned of all foreign material. Bracy Clark says that they make an audible snap in doing this, but the writer has not heard it.

Late symptoms.—The chief symptom of the disease caused by these larvæ is the catarrhal discharge on the affected side of the nose, which gives rise to one of the popular names of the disease, "snot-nose." Even this symptom may be absent when but few larvæ are present.

Neumann (*Traité des Maladies Parasitaires*, 1888, p. 501) accurately describes the symptoms of affected sheep as follows:

Three or four larvæ of *Œstrus* are frequently found in the frontal sinuses of sheep which, during life, had never manifested any symptom. It is only when these larvæ are numerous, and when they are quite well advanced in their development at the commencement of spring-time, that they occasion morbid troubles. The latter begin by a discharge, often unilateral, which is at first clear and serous, then thick and mucous. Frequently there is sneezing and snorting, accompanied by the expulsion of mucus and sometimes of *Œstrus* larvæ. Later the animals turn the head backward, often shake it, rub the nose against the ground or some other object within reach, or with their front feet. As the malady gradually advances the sheep go with lowered head, lifting the feet high as if they were walking in water. Sometimes they quickly raise the head, carrying the nose to the wind, and then bend it backward convulsively. From time to time they stagger and are seized with vertigo, but do not turn in a circle. In severer cases there is difficulty of breathing, the first respiratory passages being obstructed by the larvæ or the inflammation of the mucous membrane. The eyes are red and watery. The disease may be still further complicated. The sick lose appetite and rapidly grow poor; they grate their teeth; a frothy saliva runs from the mouth; their eyes roll in the sockets; convulsions arise and finally death ensues, sometimes within six or eight days after the appearance of the first symptoms.

But the disease is rarely so fatal; it lasts longer, and the larvæ having been successfully cast out, the symptoms generally become more favorable and by degrees completely disappear.

This affection has sometimes been mistaken for "gid," or "turn-sick," due to *Cœnurus cerebralis*, whence the name "false gid," or vertigo of *Œstrus*, which has been given to it. Confusion will be avoided by recalling that turning in a circle does not take place in the present disease. The latter is nearly always accompanied by nasal discharge and snortings, which do not appear in true "gid," and which, besides, show themselves only in young subjects.

Occurrence.—The larvæ of *Œstrus ovis* may be found in the nasal cavities throughout the year, and in nearly all stages of growth. This is more especially true of the southern portions of the United States, where the winters are mild and short. During the last winter and spring, in January and March, larvæ were collected of all sizes. Those represented by natural-sized figures (Plate I, Fig. 7), were collected in January at an abattoir in Baltimore, Md. Those figured in Plate III were collected in May. From the older of these grubs a pair of flies were hatched. The presence of very young larvæ during the past winter is very interesting, and indicates the presence of flies at an unexpected season. The usual time for the appearance of the fly is said to be during the months of June and July, and the usual period of pupation about two months. Two of my experiments showed that the time might be three weeks or four weeks exactly. The larvæ is said to dwell in the nostrils about ten months. This statement has not yet been verified. In very young lambs only young larvæ can be found, while in yearlings the larvæ may be nearly adult, depending on the exact age of the lamb and the time it was infected. In yearlings the frontal sinuses are small and the grubs easily escape detection. It is in two-year-olds and older sheep that one finds the greatest infection. In ewes the sinuses are sufficiently roomy to hold four or five larvæ without crowding, but wethers, which have small horns, or bucks which have very large frontal sinuses, can harbor many more. Rarely have more than six or seven been found. Cases have been reported in other countries where far larger numbers, as many as ten to fifteen, were found. The largest recorded number seems to be from sixty to eighty. The relatively small number discovered, and the comparatively large size of the young when deposited, indicate that each female lays but few young. It is very unusual to find more than two or three larvæ of the same size, especially if they be mature or nearly so. The young larvæ are sometimes more numerous, six or seven of nearly equal size being found together. The presence of all sizes of larvæ in the cavities is a plain contradiction to the statements made that the fly appears only in June and July, for, no matter whether it takes ten months for the larvæ to grow or not, young and middle sized and mature larvæ found in the winter time could not all have been laid within the two months indicated.

The more correct statement is that the fly may appear at any time when the temperature is not too low, but that they are more abundant in early and midsummer.

Preventive treatment.—Most authorities on this subject recommend preventive measures, but practical application of the means and remedies proposed is necessary to demonstrate their utility. A change of pasturage or an avoidance of brush-fields does not seem to be advisable unless the sheep are turned into longer grass, for the flies are able to follow the sheep wherever they may go.

A practical means of prevention consists in smearing the noses with a mixture of equal parts of tar and grease, or of tar and fish-oil, or of tar and whale-oil. The better way is to apply the preparation directly by a brush. Some recommend smearing the salt and grain troughs with the mixture, expecting the sheep while feeding to get more or less on the nose. This method is not thorough enough. Fish or whale oil alone is also recommended. Powers (American Merino, 1887, p. 300) advises the following ointment for this purpose: Beeswax, 1 pound; linseed-oil, 1 pint; carbolic acid, 4 ounces. Melt the wax and oil together, adding 2 ounces of common rosin to give body, then, as it is cooling, stir in the carbolic acid. This should be rubbed over the face and nose once in two or three days during July and August. He also recommends an apparatus which may well be used by owners of first-class breeding stock, and possibly others who own but few sheep: " A canvas face-cover smeared with this mixture (the above), or with one of asafœtida and tallow, may be hung in such fashion as not to interfere with the sight or with grazing and yet protect the lamb against the fly."

Whatever the preparation used it should be periodically repeated throughout the season during which the fly is known to trouble the sheep, as the nosing of the sheep in the grass, the accumulated dirt, and the rain all tend to make the preparation weaker and consequently less effective.

Old authorities recommend plowing furrows in the pastures, but these will be beneficial only while the ground is dry and mellow. Removal of the grub immediately after it has been deposited is impractical. All grubs seen on the ground should be crushed. Heads of slaughtered sheep should be cared for so that mature grubs can not escape to the ground. Sheep yards should be periodically cleaned and sprinkled with lime.

Medicinal treatment.—This seems to be hopeless. A study of the life history of this parasite, which appears in the south at nearly all seasons of the year, and of the anatomy of the recesses into which the larva wanders, will convince one of the difficulties to be met with in treatment. In the first place, even if a suitable remedy were found, the sheep-owner would be compelled to resort to treatment as often as he found his sheep troubled, and would have to treat each separately. This arises from the fact that irritating fumigations or sneezing powders, which pass into the lower part of the nose, would not affect the larvæ in the sinuses no matter how violent the sneezing which they excite. Injections of irritating substances would also fail, excepting possibly in the hands of an expert, who, with a syringe and peculiarly bent nozzle, could perhaps learn to inject into the nasal sinuses. Even in that case failure would result in a certain proportion of cases, and the maxillary sinuses could not be injected, nor would the larvæ in the recesses of the turbinated bones be reached. In addition to all this, most remedies which would kill the larvæ would injure the delicate mucous membranes. For the flockmaster who may wish to try fumigation or nasal injections, the following recipe for fumigation as given by Blacklock is reproduced:

One person holds the head in a convenient position in front of the operator. The latter, having half-filled a pipe with tobacco and kindled it in the usual manner, places one or two folds of a handkerchief over the opening of the bowl, then passes the stem a good way up the nostril, applies his mouth to the covered bowl, and blows vigorously through the handkerchief. When this has continued for a few seconds the pipe is withdrawn, and the operation repeated on the other nostril.

Powers (*o. c.* p. 300) advises the following nasal injection, which should succeed if any will:

It is best to procure at the drug store an elastic bulb syringe, price about $1, with a small nozzle 6 inches long. Mix turpentine and linseed-oil in equal parts. Accustom yourself to the action of the syringe so that you can gauge it accurately. Let the affected sheep be held before you in a natural position, and carefully probe the nostrils with the nozzle until you find its bearing and depth (the nozzle will pass up a surprising distance—six inches in grown sheep). Then charge the syringe, introduce it to the extremity of the nasal cavity, and with a quick pressure inject about a teaspoonful of the mixture. Withdraw at once and let the sheep recover somewhat from the effects of the shot, then treat the other nostril in the same way. * * * Keep the mixture well shaken.

If the nozzle has a properly curved tip the injected mass will be more likely to reach the larvæ. A trial on the head of a recently dead or slaughtered sheep would give the operator more knowledge of the requirements to be met than any description. Olive oil is preferable to linseed.

Surgical treatment.—There remains but one other method of removing the parasites, and that is mainly surgical. If the disease is apparent on but one side (it may be on both), an opening is made into the frontal sinus (see Plate II), with a special instrument called a trephine. The opening is made on the dotted lines which connect the middle of the eyebrows, and a little nearer the middle line of the head than the eye. The operation is a tedious one, requires some skill, and if advisable to undertake with a number of sheep, should only be trusted to a competent veterinarian. Moreover, the ultimate results are not such that the operation could be advised in the majority of cases. Yet, as it is the only means that offers any hope for the worst affected, Zürn's directions (Raillet, *Maladies Parasitaires*, p. 504) for operating are given:

Cut off the wool which covers the forehead. Trace with colored chalk a transverse line uniting the middle of the two superciliary arches (the eyebrows) and divide it by another line passing at the middle of the forehead. The point of choice for trephining will be in each of the two upper angles thus obtained without engaging the lines which limit them. The operation is performed according to the ordinary rules of surgery. From the opening made one often sees the larvæ, which are extracted by forceps. To kill others benzine moderately diluted with water is injected. The flap of skin is then cleaned, applied to the opening, and sewed to the adjoining skin. The whole is then covered with a turpentine coated leather plaster. The patient is separated for a few days from other sheep. Sheep bear the operation with the same impunity as they do marking the ears or other little operations.

Trephining may also be resorted to for the large cavities at the base of buck's horns, or for the maxillary sinus. The latter is a far more difficult operation, and the vicinity of important nerves and blood-vessels demands that only a skilled veterinarian should undertake it. The operation is, after all, only temporary in its effects, for the next larvæ laid in the nose will crawl into the same sinuses and create the same disturbances as those removed.

Neumann's advice, with which he closes his chapter on this parasite, is, perhaps, the soundest to follow, except in the case of breeders of valuable sheep: •

At all times, if the number of animals affected is considerable, the malady should be left to follow its course, and those which present the gravest symptoms should be sent to the shambles.

ŒSTRUS OVIS, Linn.

PLATE.I

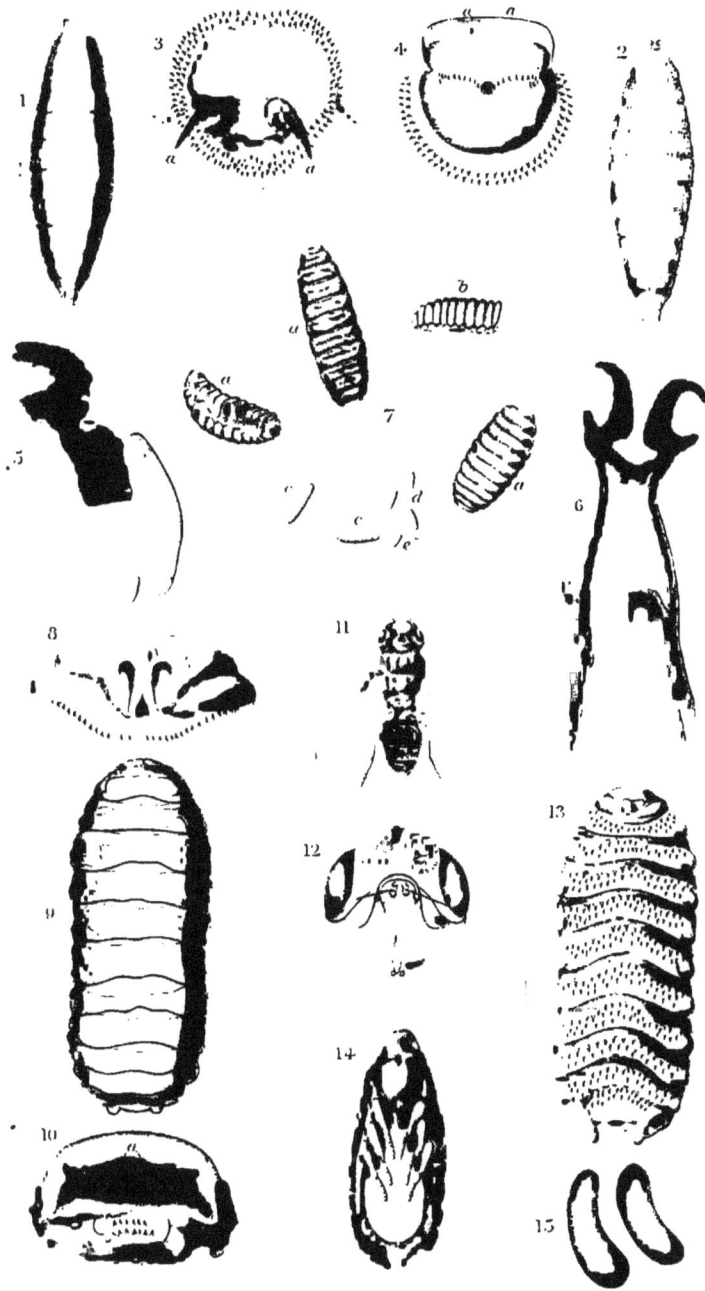

Haines, del.

A.Hoen & Co. Lith. Baltimore

ŒSTRUS OVIS,
(The Gad Fly of Sheep.)

ŒSTRUS OVIS, Linn.

PLATE II.

Dissection of the head of a sheep to show the cavities into which the gad-fly grubs penetrate. The straws indicate the passages from the cavities into the inside of the nose: *a*, the superior maxillary sinus in which there is imprisoned the calcified remains of a fully developed larva; *b*, the ridge made by the infra-orbital division of the fifth pair of nerves; *c*, the infra-orbital foramen; *d*, an opening into the nares; *e*, *e*, the frontal sinuses with young larvæ in them. The dotted line *f*, *f*, indicates the level at which trephining, if it is done, should be performed. Figure reduced to three-fourths of natural size.

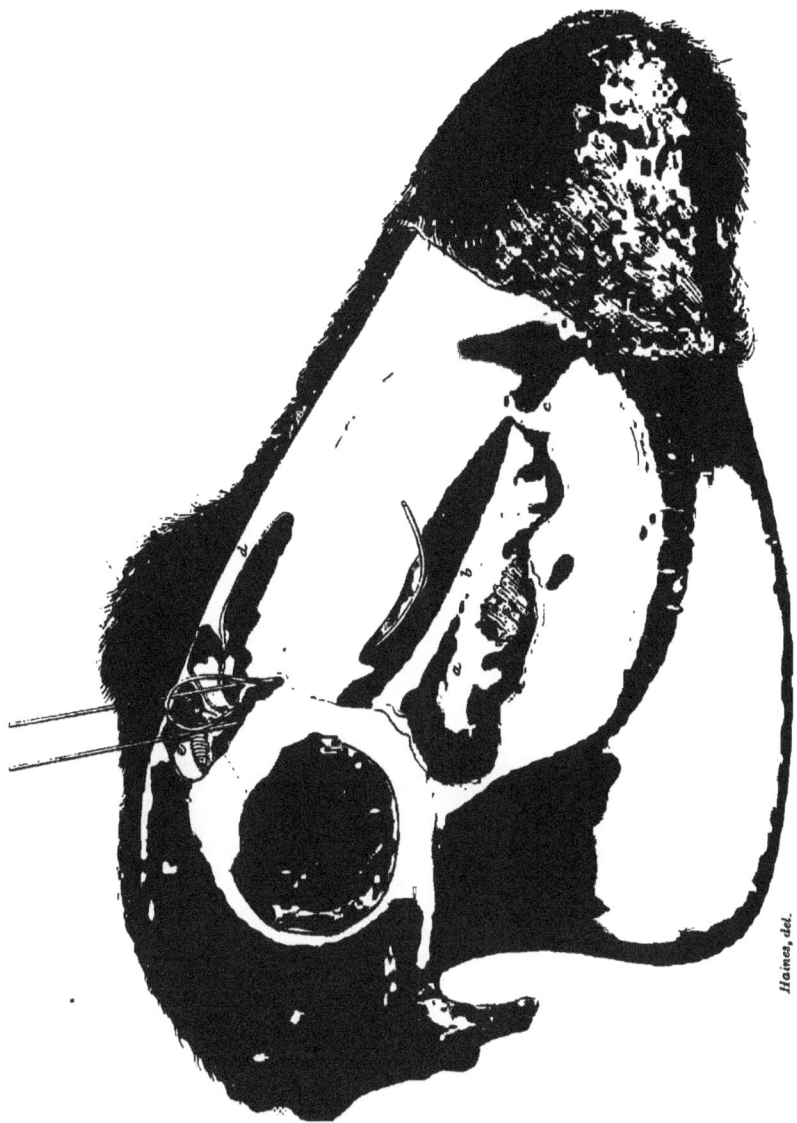

Haines, del.

ŒSTRUS OVIS, Linn.

PLATE III.

Fig. 1. Section of head of sheep made a little to the right of mesial plane: *a, a, a, a, a,* section of bone surrounding *b,* the brain, and *c,* the nasial cavity; *d,* the lower jaw bone; *e,* nostril; *f,* opening of tear duct; *g, g, g,* turbinated bones; *h,* the posterior opening of the nasal cavity, and near the opening of Eustachian tube; *i,* placed on the turbinated bone over a grub in the groove leading to the frontal sinus; *k,* the frontal sinus; *l,* the nasal sinus; *m,* the thin perforated plate of bone called the cribriform plate; *n, n, n,* larvæ of *Œstrus* ascending the nares and wandering about its surface; *i,* one ascending to the sinus; *k,* opposite three, nearly mature larvæ in the sinus. Figure reduced to three-fourths of natural size.

Fig. 2. Outline drawing of the skull surrounding the frontal sinus after removal of part of the turbinated bone, lettered as in Fig. 1; *o, o,* straws passed through the channels connecting the nares with the sinuses, marking the path by which the larva reaches the sinus; *p,* cut ends of the removed bones.

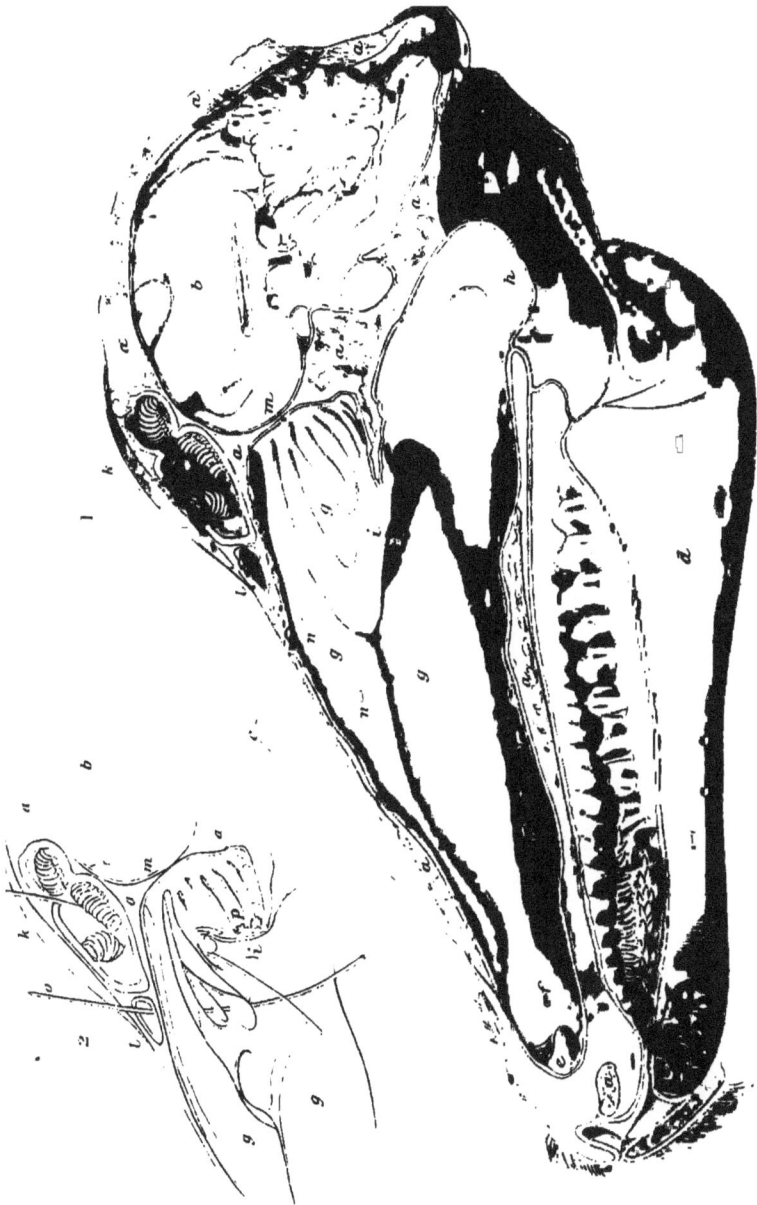

THE SHEEP-TICK OR LOUSE-FLY--PHTHIRIASIS.

MELOPHAGUS OVINUS, Linn.

Plate IV.

One of the best known of all the external parasites of sheep is the sheep-tick, *Melophagus ovinus*, Linn. This pest is a very common one in the Eastern States, and although it seldom causes any serious damage either to the sheep or to the wool, it is at all times an annoyance, and occasionally causes decided losses to the flock-master.

This tick, like a majority of the parasites of the domesticated animals, was introduced into this country from Europe. The name " sheep-tick," though not a misnomer, as every one knows what a sheep-tick is, is misleading. So much do they resemble the true ticks that they are often classified together. The most superficial study will, however, serve to show their differences.

Description.—The sheep-tick is a wingless fly having but six legs, whereas the true ticks are more closely related to spiders, and have eight legs in their adult state. The adults are less than a quarter of an inch long, and have a short, flattened, bristly, leathery body. The head is slightly wider than the thorax, into which it is sunk. They have very short antennæ, which are sunk in sockets in the face ; the proboscis is tubular, and is protected externally by two flat, elongated bristly pieces, the labrum ; its end is armed with teeth. The thorax or limb-bearing portion of their bodies is nearly square when looked at from above. It is composed of three pieces, the middle being the one seen on the back. The legs are very stout, covered with bristles, and each is provided with two strong, sharp claws. The last joint of each foot bears a pinniform or feather-like organ whose office is as yet undetermined, but is probably that of coiling around hair for better prehension. There are no wings. On either side may be seen two small, bristle-covered, round spots at points where the wings should be attached. At the posterior outer corners of the thorax are two little projections which remind one of balancers. The abdomen, usually larger in females than in males, is flattened and bag-like, and is as large or larger than the rest of the body, especially after the louse-fly has eaten, when the red blood sucked from its host may be seen through its skin. Its skin is tough, unsegmented, semi-translucent, and permits the abdominal organs to show through. On each side there are seven stigmata or breathing pores. The anus is situated on the ventral side and just behind the genital orifice. The sexes resemble each other, but may be separated by their size and by the form of the external genital apparatus.

In habit these parasites resemble lice living among the hairs of the fleece, whence the name, louse-fly. They seldom remain attached to the skin longer than a sufficient time to fill up with blood; this they suck up through the proboscis with which they perforate the skin. They try to evade capture by running into the wool, and when caught cling tenaciously.

Life history.—The family of flies to which this parasite belongs is truly wonderful, in that they bring forth their young as puparia. The puparia of *Melophagus* are laid as flat, ovoid, chestnut brown, glistening seed-like eggs, which are nearly one-third as large as the abdomen of the parent, and contain an imperfectly developed larva within them. The egg cases or pupa are marked by two rows of seven dots each on one surface, a slight depression, indented by two dots in one end, and a slight elevation at the other. The two dots are at the anal end. A dissection of one of these, taken from a female, shows the pointed end to be connected with a membrane, and to be the end through which the larva obtains food.

Some authorities say that each female produces but one or two of these puparia; others say that they can lay about eight or nine during their life-time. They lay one at a time in the wool. A portion of each puparium will be found to be covered with a dry, dark substance, which came from the parent when the puparium was laid, was sticky, and glued it to the surrounding hairs. This prevents the pupa, which becomes hard, dry, and glassy, falling from the wool. The insect emerge, with adult characters from the pupæ cases, through an opening in the end of the case made by a round lid splitting off, and wanders into the wool. Some of these eggs collected in the course of our investigations hatched within four weeks at ordinary temperatures.

Occurrence.—These ticks, or their young, may be found on the sheep at all times of the year, but appear to be most numerous in spring. They are particularly noticeable at shearing time on the old sheep after they are deprived of their shelter. At this time those that can do so take refuge in the longer wool of the lambs, and prove veritable pests. The others perish either from being eaten by the sheep, carried away in the wool, or dropped to the ground. They frequently become attached to the clothing and persons of people with whom they come in contact, but they prove but a slight source of discomfort, as they are easily caught and killed. It is not at all probable that they can exist many days apart from the sheep, as they are unfitted by structure for any other habitat. Their food consists wholly of the blood which they suck from the sheep. They depend also on the sheep for warmth; in warm spring days they may be found crawling near the ends of the wool, while in the colder days they will always be found either engaged in feeding or resting at but little distance from the roots of the wool. A dozen or more of these ticks, which were moderately well fed when taken from the sheep were, with some wool, placed in a cotton-stoppered bottle and kept in a room with a temperature varying between 60° and 80° Fah. They all died in less than four days. The leanest succumbed first, in about two days, while those that were better nourished gradually grew smaller and thinner, and lived little longer than the third day. Others placed in wool over the damp soil of a geranium in a flower-pot died within four days. On the other hand, some young ones which were hatched out in a bottle were kept for nearly two weeks, or until their daily feeding was neglected. To feed them they were placed on the back of my hand. By this means I could, with a lens, watch them bore into the skin and see the abdomen slowly enlarge as they drew in the blood. They had some difficulty in penetrating the thick skin, but usually succeeded by slipping their tube into a hair follicle. At first no itching or irritation was felt, except a slight twinge when the bills first penetrated the skin ; but little swellings came on a day or two later which itched for over a week. The itching was far more persistent than with mosquito bites. They must in this way cause lambs much discomfort. It was at first thought that a fluid could be seen running from the parasite to the hand through the bill, but no more was thought of it until the elevations began to rise and itch ; then it appeared certain that the little pests had secreted a poisonous fluid. The office of this secreted fluid is probably to assist the flow of blood by keeping it from clotting.

The above experiments show that the parasite spends its whole life on the sheep.

Source of contagion.—The fact that this parasite passes its whole life on the sheep, and that it produces but few young, are very important considerations in efforts for exterminating the pest and preventing a new infection. They indicate that if all are killed the sheep will not again be infected except from other sheep.

Disease.—The injury sustained by sheep from these parasites varies according to the number present. It arises from the itching and pain inflicted by them when obtaining their food. A few cause but little annoyance, and, were it not that these may become the source of future multitudes, would be scarcely worth noticing. To lambs the annoyance is particularly aggravating, as their skin is tender and the number of parasites attacking them after shearing is unusually large. In older sheep the irritation is the more noticeable towards spring. At this time the parasites are more numerous and the animals bite and scratch themselves oftener.

Medical treatment.—The well-known means of ridding the sheep of these pests are the tobacco or other mixtures used for scab-dips. The best time of the year to dip is at shearing time. Then the older sheep can easily be handled and cleansed. The dipping should not be delayed long after, for each additional day is one of torment to the lambs. The lambs also should be thoroughly dipped. One dip is said to be sufficient to kill the old parasites. However, a few of the pupæ may remain in the fleece of the lambs. They should be thoroughly examined two or three weeks after, and if there are any present they should again be dipped. All the precautions taken in dipping sheep should be carefully observed. The shorn wool should be stored where the young ticks which may hatch from the pupa cases can not crawl back to the sheep. However, they seem to have great difficulty in crawling, and may not be able to go far. An experiment in which some of the pupæ became wet with moisture from other ticks in the same bottle demonstrated that they would not hatch, and indicates that the sheep bath will very likely kill the inclosed larvæ.

In *The American Agriculturist*, October, 1889, page 490, Mr. Joseph Harris advises fall dipping for these pests. This is a good plan if the sheep have become infested with ticks after the spring dipping by some inadvertence of the master. He advises the use of tobacco, carbolic acid, and kerosene emulsion dips. The carbolic acid dip is composed of a pound of soap and a pint of crude carbolic acid to each 50 gallons of water. Dissolve the soap in a gallon or more of boiling water, add the acid and stir thoroughly. Keep the mixture well thinned, and do not let it get into the mouths, nostrils, or eyes of the sheep. Hold each sheep in the bath not less than half a minute.

The formula for kerosene emulsion is as follows: Churn fresh skimmed milk and kerosene together in the proportion of 1 gallon of milk to 2 gallons of kerosene, either in a churn or by using a force-pump until an emulsion is made. The method of using the force-pump is to set it in the vessel containing the mixture and turn the stream back into the same vessel. The emulsion will form quicker if boiling hot milk is used. For dipping use 1 gallon of the emulsion for each 10 gallons of water required. Mr. Harris seems to think 20 gallons, with a reserve of 10 gallons, sufficient; but he was evidently thinking of a very few sheep. He did not use this emulsion, but a variation made with soap, as follows:

Boil a gallon of water, dissolving a pound of soap in it; add 2 gallons kerosene; churn the mixture until it emulsifies, or until all the oil is "cut." Use 1 gallon of emulsion to 8 of water. Mr. Harris advises dipping twice with an interim of two weeks.

Fifty gallons of the dip will suffice for fifty sheep. Seventy would probably answer for one hundred; but much depends upon the amount of waste on account of the liquor being carried off by the fleece. The recipe is very easily modified for ranching purposes.

The emulsion has already been advised for cattle lice in Bulletin No. 5, Iowa Agricultural Station, p. 184, May, 1889, and for cattle ticks in *Insect Life*, Vol. II, No. 1, p. 20, U. S. Department of Agriculture, July, 1889. Though the efficacy of this remedy against the various kinds of insect parasites of domestic animals has yet to be tried in detail, its importance in this field has already been demonstrated.

Preventive treatment.—The sheep should not be turned into the old pens or pastures until a week after the first dip, by which time it may reasonably be supposed that all parasites on the ground have died. To avoid the chance of any recently hatched parasites getting from the ground to sheep in places where the sheep rest, it is best to scrape out and cleanse the pens. Animals recently purchased should always be dipped before being added to the older flock.

By following out a thorough plan of treatment, and by carefully guarding the sheep from re-infection, the flockmaster should be able to rid his flocks of this pest in a single season.

MELOPHAGUS OVINUS, Linn.

PLATE IV.

Fig. 1. Female sheep-tick : 1a, larva case, each natural size.

Fig. 2. Male, dorsal view, ×8: a, head; b, thorax; c, abdomen; d, limbs; e, oval bristle-covered disks, which correspond to the points of attachment of wings in other flies; f, rudimentary halteres or poisers.

Fig. 3. Male, ventral view, ×8: g, h, and i, the three segments of the thorax; k, the external genitals.

Fig. 4. Female, dorsal view, ×8.

Fig. 5. Larva case, ×8: a, cephalic end ; b, two rows each of seven shallow indentations.

Fig. 6. Foot: a, the two claws between which hangs b, the pinniform prehensile organ ; c, the tarsi, whose last joint d supports the prehensile organ ; e, distal end of the tibia.

Fig. 6a. Prehensile organ, ×60 : a, the segmented muscular portion included within the tarsus; b, the flexible grasping portion.

Fig. 7. Front view of head : a, the compound eyes; b, the antennæ sunk in cuplike cavities; c, the labrum which protects the sucking organ.

Fig. 8. The sucking apparatus: a, the labrum ; b, the orifice from which the tube protrudes; c, the sucking tube.

Fig. 9. End of the sucking tube, ×120: a, teeth by which the tube cuts its way through the skin ; b, rod upon which the teeth are set ; c, tube which has lateral orifice in it. Other details not shown.

Fig. 10. External genital apparatus of female: a, spine-covered cap which fits over b, the genital orifice above; c, two clusters of spines which seem to be for clasping; d, the terminal of the seven pair of stigmata or breathing pores ; e, anus.

Fig. 11. External genital apparatus of male: a, the two lateral of the three chitinous styles which surround the projecting intromittent organ ; b, two clusters of spines which seem to be claspers; c, stigmata.

Fig. 12. Larva case, ×6: a, case with the broken operculum inside ; b, cephalic end, showing line where the operculum splits off and the remains of a central orifice through which nourishment was obtained by the embryo from the parent ; c, caudal end showing the two dots corresponding to the two terminal stigmata.

Fig. 13. Larva case with larva, ×6: a, ventral view ; b, dorsal view.

PLATE IV

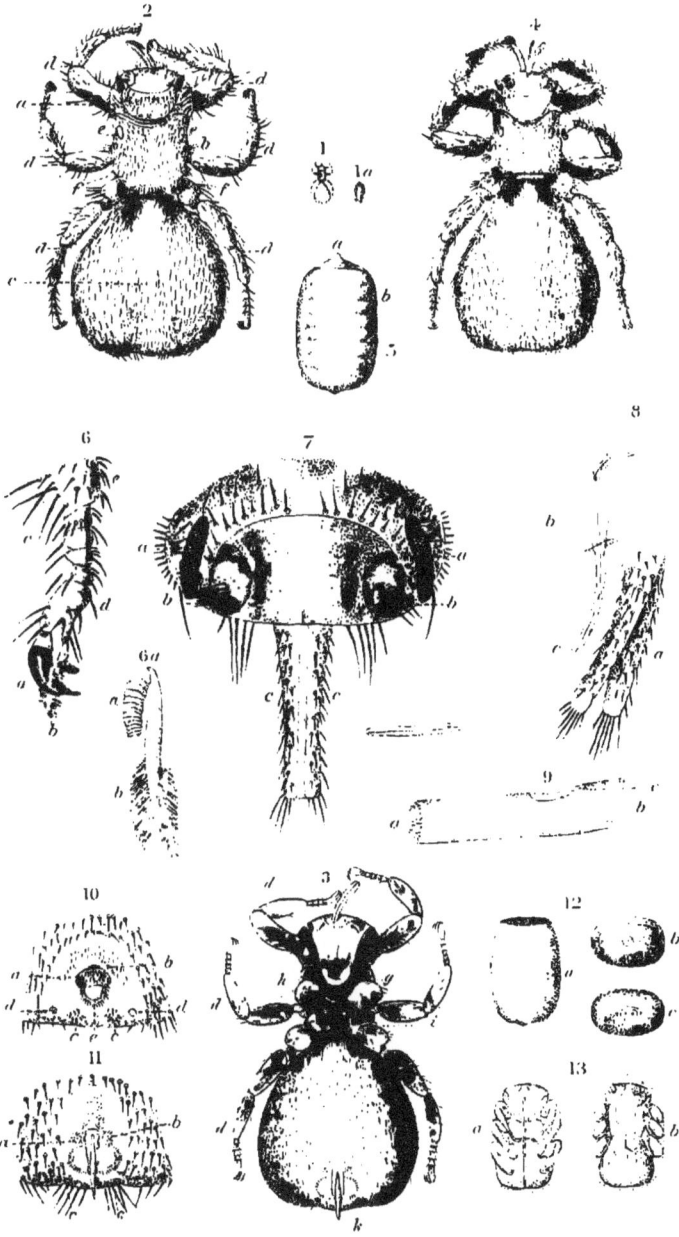

MELOPHAGUS OVINUS.
(The Louse-fly.)

THE SHEEP-LOUSE—PHTHIRIASIS—LOUSINESS.

TRICHODECTES SPHÆROCEPHALUS, Nitzsch.

Plate V.

The little red-headed sheep-louse is not very abundant in this country, and easily escapes detection. It is, nevertheless, of sufficient economic importance to engage our attention.

The genus *Trichodectes*, to which this parasite belongs, is classed by scientists in the order *Mallophaga*, to which order the bird-lice belong. They are said by Packard (*Guide to Study of Insects*, p. 554), to live upon the hairs of mammalia and feathers of birds.

Description.—The sheep-louse is quite small, about 1ᵐᵐ or one-twenty-fifth of an inch in length. The female is slightly longer and larger. It is characterized by its reddish head and the pale, transverse bands which cross the abdomen. Neumann describes the species as follows:

> Its head is wider than long; truncated in front, the antennal band making the turn of the forehead which carries long hairs on its border. The antennæ are hairy and a little longer in the male than in the female. The abdomen carries sub-quadrangular median spots. The general color is whitish; the spots and head are ferruginous. Length of female 1.6ᵐᵐ, and male 1.4ᵐᵐ.

This general description is very good, but would hardly serve to differentiate this species from others without figures and comparative descriptions. Entomological anatomists enter into very minute details of description for determining these species, but a comparison of specimens found with the figures in Plates V and VI will assist the farmer more than a long technical description. As yet the only form recorded from sheep in this country is the one species, *Trichodectes sphærocephalus*, or *T. ovis*, which is a synonym. The following appear to be the most marked differences between this and other species: The species is generally smaller than others. The breadth of the abdomen is relatively narrower, and that of the male seems to be more obtuse. The dorsal sutures on the head (see Plate V, Fig. 2 *e, e*), are fainter in the middle. The front of the head is more convex. The brown markings on the head are all relatively fainter; those on the back of the male and female increase from the first to the fourth and then decrease to the last in a gradual manner. The claws of both anterior and posterior feet are more nearly of the same size. The eggs (Figs. 6 and 7) are midway in size between those of the two species figured on Plate VI. These eggs have a cap marked by vertical striæ, which constitute a generic charac-

45

ter. The markings on the under side of the head and the form of the caudal end of male and female seem to differ from others, but these differences are unessential for the present description. The form that most nearly approaches it in general appearance is the *Trichodectes pilosus*, Giebel, from the horse, but the latter is a relatively larger species. The species is to be found on poorly nourished young sheep in places where the wool is scanty. A favorite place is between the legs and body, just under the shoulder. Coarse-wooled sheep are more afflicted with them.

The life history of these pests is very simple. The adults lay their eggs on the wool fibers at their base, and a glutinous material sticks them there (Plate V, Fig. 6). The eggs hatch out in the wool, and the young louse emerges by pushing off the cap (Figs. 7 and 3). The young then grow to be adult. It is most probable that sheep can only get the lice from other sheep, as another host of *Trichodectes sphærocephalus* is yet unknown. The presence of these parasites may easily be learned by searching for them or their eggs. Sheep affected manifest their presence by scratching themselves with their hind feet or by rubbing against stationary objects.

Disease.—The injury wrought by this parasite is caused by its life-habits. The injury effected by species of *Trichodectes* is not as decided as that caused by those of *Hæmatopinus*, the genus to which the true lice belong. The mouth parts of the latter are so arranged that it can live on the blood of its host, and in biting through the skin it causes an itching sensation and a wound. The *Trichodectes*, however, are not fitted for penetrating so deeply, and appear to go but little deeper than the epithelium. They can probably bite through to the young growing tissue, for animals which are severely afflicted with these pests have a roughened, scabby skin, which would not be the case if the parasite only lived on the hairs and epithelial débris. The presence of these parasites on the skin not only gives discomfort to the sheep, but causes the skin to thicken, become rough and covered with little dry, black scabs, and the wool to become short, dry, gnarly and worthless wherever the pests attack the skin,

TRICHODECTES SPHÆROCEPHALUS, N.

PLATE V.

In Plate V, figs. 1, 2, 3, 6, and 7, and in Plate VI, figs. 1, 2, 3, 5, 11, 12, 13, and 15 are equally enlarged, and present relative differences in size and form. Other parts are also enlarged similarly for the sake of comparison.

Fig. 1. Male, natural length indicated by line.

Fig. 2. Female, natural length indicated by line: *a*, head; *b*, antennæ; *c*, face; *d*, cheeks; *e*, *e*, dorsal sutures; *f*,*f*,*f*, legs; *g*, prothorax; *h*, metathorax; *i*, abdomen; *k*, dark transverse bands; *l*, line of hairs on each segment; *m*, *m*, stigmata or breathing pores; *n* (fig. 1), male genital orifice; *o*, female genital orifice; *p*, female claspers.

Fig. 3. Young specimen just emerged from shell.

Fig. 4. Male antenna.

Fig. 5. Female antenna.

Fig. 6. Egg soon after being laid : *a*, cap with peculiar rod-like structure; *b*, line at which the cap is to cleave off.

Fig. 7. Egg shell which has lost its embryo and cap: *b*, *b*, wool fibers.

Fig. 8. Anterior leg: *a*, coxa; *b*, trochanter; *c*, femur; *d*, tibia; *e*, tarsi and claws.

Fig. 9. Posterior leg.

Fig. 10. Head, ventral side: *a*, *a*, antenna; *b*, *b*, ventral continuation of dorsal suture; *c*, *c*, ventral suture; *d*,*d*, mandibles; *e*, maxillæ showing through the chitinous gular plate; *f*, the labrum.

Fig. 11. Tail end of male, dorsal view: *a*, the last segment; *b*, the genital orifice; *c*, chitinous, hook-like appendages of the genital apparatus.

Fig. 12. Tail end of female, ventral view: *a*, the last segment; *b*, the genital and anal orifice ; *c*, the claspers.

PLATE V

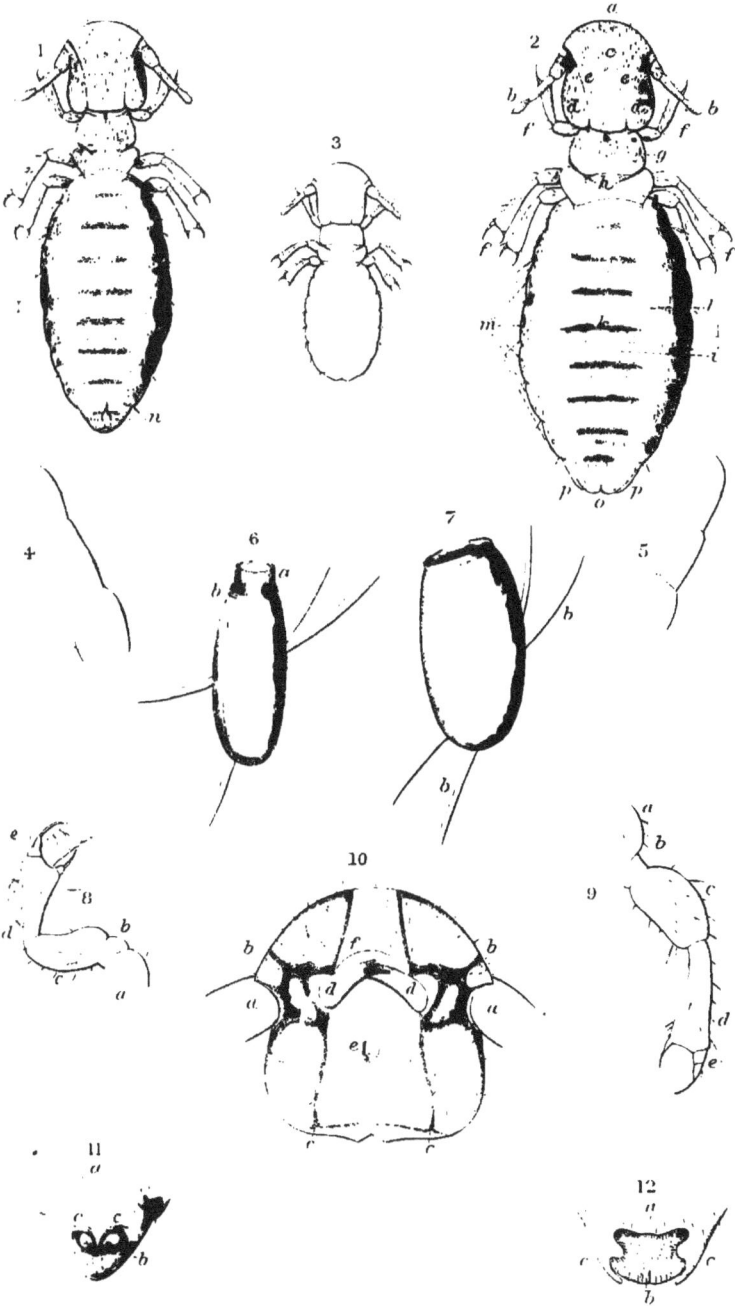

TRICHODECTES SPHÆROCEPHALUS,
(The Sheep Louse.)

· GOAT LICE.

Description.—*Trichodectes limbatus*, Gervais—the Angora louse—resembles *T. climax* closely, but differs in specific details. All the brown markings on *T. limbatus* are darker and wider; the head is slightly more indented in front; the dorsal suture (Plate VI, fig. 2 *e, e,*) is more pronounced; the head is wider in proportion to its length. The banded margin of the abdomen is wider. The greatest difference lies in the disproportion of size between the male and female of *T. limbatus,* and the character of the transverse bands of the abdomen of the male; the abdomen of the male being shorter is relatively wider in proportion to its length, and has a quadrangular appearance. The first band is nearly straight and one-third shorter than the three succeeding, which are of nearly equal length and convex anteriorly, concave posteriorly. The fourth and fifth segments bear a second narrow band near the posterior margin. The egg of this species is larger than that of *T. sphærocephalus* or *T. climax.*

The marked differences shown between the bands of *T. climax* and *T. limbatus* was a constant one in all males examined. This feature, in connection with the difference in size of the eggs and the many minor differences of form and color, seem to be specific. The males of these species seem to offer the most tangible characters for separation. Whenever other species of this genus are described particular attention to the males should, on this account, be given. The females seem to approach each other more closely.

Occurrence and disease.—The goat louse is common, and causes more trouble to goats than the little red-headed louse does to sheep. When present it occurs among the coarse hair along the back and sides of the goat. It causes much discomfort and scabbiness of skin. If the animals are severely infected they become poor and thin. The Angora goat louse caused, in the single flock in which it was seen, not only a severe scabby skin disease, but a loss of fleece. The whole back, sides, and head seemed to be completely covered with the parasites, young and old, and nits.

Treatment.—The disease caused by these parasites is of that class which is preventible, and it is inexcusable if it be allowed to continue. The history in each case is that of infection from other sheep, goats, or Angoras, as the case may be. As the parasites spend their whole lives on these animals they may be killed on them and not be expected to come again except from other animals of the same kind. Yards, where any of these animals are kept, should be sprinkled with lime and the walls washed with diluted lye, or whitewashed before the animals which have been treated are returned. This will ensure safety from any parasites which may have fallen to the ground where the goats have lain. The treatment should be thorough. As these animals are small the best method is to immerse them in tobacco water, thus insuring the destruction of every parasite and nit, even those on the nose, by immersing them while holding the nostrils. If the hair is long on the Angora they should by all means be sheared. If it is not desirable to dip them, the selected remedy may be sopped on the skin and wool and thoroughly rubbed in so as to wet the skin.

Medicines may be applied in three forms: in powders, as pyrethrum or Persian insect powder, and tobacco dust; in ointments, as oil or lard, with some added ingredient, and in baths, as the tobacco or arsenical dips. Of these the first is the more objectionable and the least valuable. The second is better, but not entirely successful. The third is the one which should be used in the majority of cases, as it is the most certain.

The following recipes are among those in use, and are recommended by various authorities :

A decoction of stavesacre seeds, 3 ounces to 2 quarts of water, to be thoroughly rubbed in. (Neumann.)

A decoction of stavesacre seeds, 1 ounce to 1 quart of water or vinegar, or half water and half vinegar. (Finlay Dun.)

In powders, tobacco, pyrethrum, stavesacre and sabadilla may be used, but the last two are not advisable.

Benzine 1 part, soft soap 6 parts, water 20 parts, or petroleum (kerosene) 1 part, sweet oil 10 parts. (Neumann.)

Schlegs' mixture is recommended in Germany for its efficacy and harmlessness when prudently used. Arsenious acid one-half ounce, potash one-half ounce, water 3 pints, vinegar 3 pints. (Zürn.)

The most efficacious remedies are the tobacco, or the tobacco and sulphur dips, advised for scab. For flocks of large numbers, nearly all other recipes are unavailable.

Mercurial salves should not be used.

Zürn advises tobacco 1 part, water 20 parts, or water 20 parts and vinegar 10 parts, to be made into a decoction, and vinegar added after cooling. The kerosene emulsion may also prove a valuable remedy. It should be applied as directed for exterminating sheep-ticks, or by means, of a force-pump and spray-nozzle.

TRICHODECTES CLIMAX, Nitzsch.

Plate VI, Figs. 11–18.

The common goat, *Capra hircus*, L., is quite commonly infested by a species of louse which has been identified by the writer as *Trichodectes climax*, Nitzsch, and the Angora goat, *Capra hircus*, var. *Angorensis*, by one which seems to be *Trichodectes limbatus*, Gervais, or *T. climax*, var. *major*, Piaget. The name applied by Gervais is accepted, for the differences between the two appear to be specific and not varietal.

Description.—The characters of *Trichodectes climax* are : Head wider than long, quadrangular, presenting a wide but shallow indentation in front, at which the two antennal bands stop; antennæ hairy, a little longer with the male than the female; the first joint is larger and shorter than the others; the second longer than the third; the abdomen bears median spots, the width of which diminish as their length increases. The last segment of the male carries two hairy cushions. Head and thorax, reddish-brown; abdomen, pale yellow; spots, brownmaroon; bands, blackened. Length, female, 1.6^{mm}; male, 1.3^{mm}. (Neumann.)

The female of this species is broader and the male a little shorter than in the corresponding sexes of *T. sphærocephalus*. The dorsal sutures of the head are much darker and plainer; the edge of the head and abdomen are margined by a pronounced dark band. The differences between the anterior and posterior feet are much more decided. The dark bands of the back of the abdomen of the male seem to offer the best characteristics, viz : They gradually increase in length to the fourth, when they begin to narrow. Beginning next the thorax the first band is narrow; the second, third, and fourth are wider; the succeeding are narrower; posterior to the line of hairs on the second to the fifth segments are extra narrow bands, which are about equal in width. The egg of this species is shorter and narrower.

TRICHODECTES LIMBATUS, Gervais.

PLATE VI.

Fig. 1. Male, natural length indicated by line.

Fig. 2. Female, natural length indicated by line: *a*, head; *b*, antennæ; *c*, clypeus; *d*, cheeks; *e, e*, dorsal sutures; *f, f, f*, legs; *g*, prothorax; *h*, metathorax; *i*, abdomen; *k, k*, dark transverse bands; *l, l*, lines of hairs; *m, m*, breathing pores; *n*, male genital orifice; *o*, female genital orifice; *p*, female claspers; *q*, male genital hooks.

Fig. 3. Head, ventral view: *a*, antennæ; *b*, mandibles.

Fig. 4. Posterior end of female, ventral view: *a*, genital and anal orifice; *b*, claspers.

Fig. 5. Egg: *a*, the cap; *b*, the line where it splits off.

Fig. 6. Antenna of female.

Fig. 7. Anterior leg: *a*, coxa; *b*, trochanter; *c*, femur; *d*, tibia; *e*, tarsi and claws.

Fig. 8. Posterior leg.

Trichodectes climax, N.

Figs. 11 to 18. Numbered and lettered for the same parts as Fig. 1–8.

PLATE VI

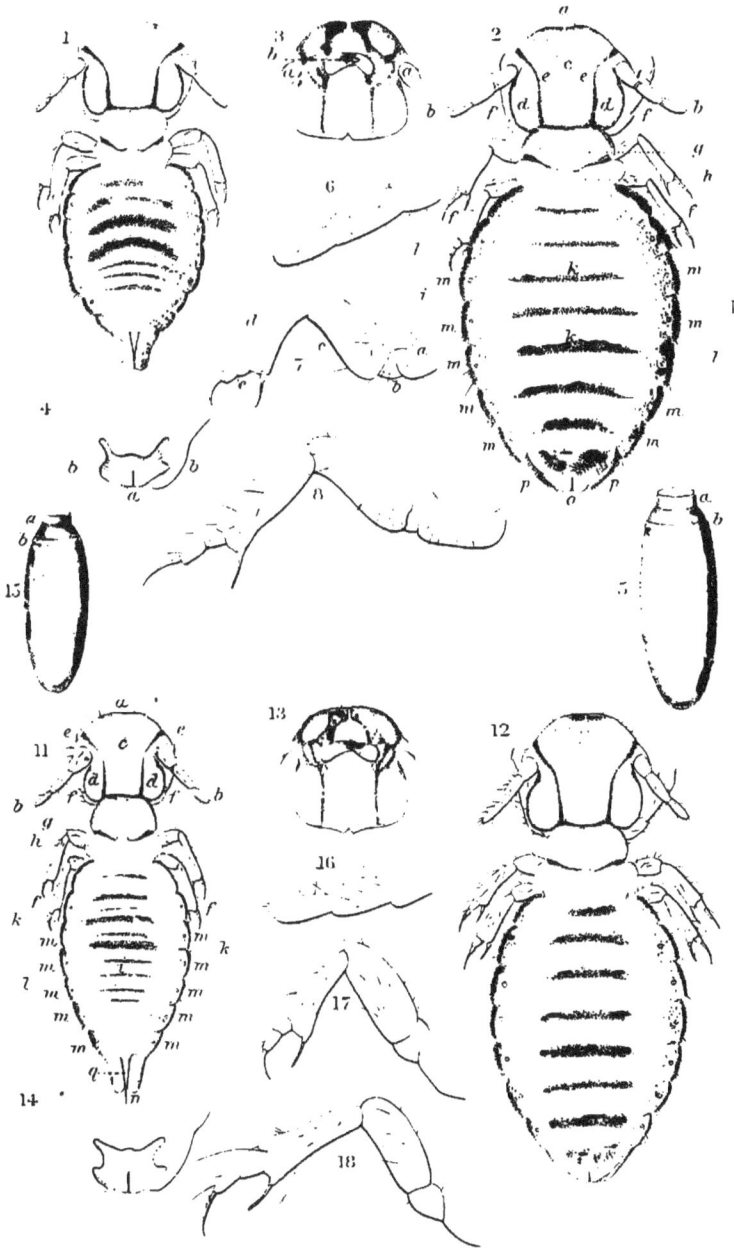

A. Hoen & Co. Lith. Baltimore

TRICHODECTES LIMBATUS,
(The Angora Goat Louse.)

TRICHODECTES CLIMAX,
(The Common Goat Louse.)

THE ITCH OR SCAB INSECTS—ACARIASIS—ITCH—SCAB.

SARCOPTES, Latr.; PSOROPTES, P. Gerv.; CHORIOPTES, P. Gerv.

Of all the diseases caused by external parasites those due to the scab-making insects are of the first importance. The losses due to them are very heavy, and are exceeded by those of no other external parasite, and equal those of the most destructive of internal parasites. Owing to the means used in preventing flocks from becoming infected, and to the extermination of the pests by the proper use of efficacious remedies, these losses are annually becoming reduced. It is to be hoped that in the near future, through the exercise of proper sanitary laws, this most tractable of all the parasitic diseases may be completely exterminated in our country.

Scab is a disease due to the presence of minute insects which lead a parasitic life on the skin of their hosts. It is caused by the inflammation they excite in penetrating the skin, that they may procure food for themselves and young, and suitable conditions under which the eggs may be deposited and hatched and the young matured. The disease is thought by some to be aggravated by a poisonous fluid secreted by the insects, which adds to the intense itching.

The malady proceeds step by step with the growth, propagation and decay of the innumerable insects which result from the acquisition of a single fertile female or a few pairs of individuals, and the spread of the disease, therefore, coincides with an increase in the numbers of the parasites, while the limitation of the disease follows their extermination.

There are at least three different species of scab-forming insects, parasitic on sheep, and each species is not only specifically different from the other, but the disease caused by each is different. This is due to the variation in the anatomical structure and habits of life in the several species of these pests, which causes them to attack the skin in different places and by different methods, and hence gives rise to the various symptoms common to each disease.

These insects are known as *Acari*, and the diseases they cause as *acariasis*. The various species parasitic on sheep are *Sarcoptes scabiei*, de Geer, var. *ovis; Psoroptes communis*, Fürst, var. *ovis; Chorioptes communis*, Verheyen, var. *ovis*. Of these the *Sarcoptes* causes scab of the head; *Psoroptes* causes common scab, and the *Chorioptes* foot scab.

The life history of these parasites is in general very similar. They attack the external skin of the animals in which they live by biting it. Soon after scabs are formed. Under these scabs the insects lay their ovoid eggs. After two or three days these eggs hatch, and in fifteen days the progeny become adult. Each adult female is estimated to lay about fifteen eggs, two-thirds of which produce females. When hatched the young invade new territory and lead the life of their parents. The extension of the invading hosts is due to their migration and rapid propagation. The disease usually spreads as a constantly growing patch. The infected sheep sometimes scatter the scabs by scratching; these, in turn, become new centers of infection. The multiplication of the invading pests ceases only at the death of the host, or when they are killed by the use of proper remedies. To illustrate the rapid increase of the *Sarcoptes*, Gerlach, a scientist, computed that in three months a single female would produce 1,500,000 progeny. He estimated that each adult female laid fifteen eggs, of which ten were females, and that the eggs became adults in fifteen days. The result is shown in tabular form as follows:

						Females.	Males.
First generation after 15 days produces					....	10	5
Second	"	"	30	"	"	....	100	50
Third	"	"	45	"	"	....	1,000	500
Fourth	"	"	60	"	"	....	10,000	5,000
Fifth	"	"	75	"	"	....	100,000	50,000
Sixth	"	"	90	"	"	1,000,000	500,000

This table presents a very moderate estimate of the rate of propagation. A microscopic examination of minute particles of scab shows them to be teeming with young and old parasites, and would seem to confirm the estimate given. As but few of the parasites may be transferred to a healthy animal, it is evident that up to the second month but little advance in the disease will be noticed, but after that time the tenfold increase every two weeks produces an enormous number of the parasites and causes the disease to advance with wonderful rapidity.

THE HEAD SCAB.

Sarcoptes scabiei, de Geer, var. ovis.

Description.—The insects which cause this variety of scab are almost invisible to the unaided eye, and are among the smallest of the scab-making insects. They are known as *Sarcoptes scabiei*, de Geer, var. *ovis*. They may be recognized by their rounded or somewhat oval bodies, their small heads, which are furnished with a biting apparatus, and by the adult having four pairs of legs. The young have but three pairs of legs. Other anatomical characters, which are different in different species, are present, but for the flockmaster these are unessential, as the species can be separated by biological characters which are plainer and easily understood.

Disease.—Head scab begins on the upper lip, and about the nostrils; more rarely it may show itself for the first time about the eyelids and ears. In these places there is less hair and grease, affording the pests better opportunities of getting at the skin. From these starting-points the scabs spread over the forehead, cheeks, eyelids, and occasionally over the space under the jaw. In badly infected sheep the disease may sometimes extend to the fore limbs, under the belly, around the joints, and especially between the folds of the knees, hocks, and pasterns. Sheep with coarse dry wool are more likely to suffer this extended invasion than those with fine, oily, and soft wool. Long wool seems to offer a barrier to its progress, for the invasion of parts covered by short wool is much more rapid. The demarkation between the invaded parts of the head and the healthy wool-bearing portions is quite abrupt. In coarse-wooled breeds the disease may rarely cover the entire body. The first indication of the disease is shown by the sheep in rubbing or scratching its head. The intensity of the itching is manifested by the violence of the sheep's action. The first that can be seen on an infected spot is little elevations with soft centers. These elevations break of themselves, or through the rubbing they receive, and from them runs a watery fluid that in drying forms little hard bunches which stick to the skin and adjacent hairs. These little elevations are made by the parasites, which sink themselves into the skin. Here the parasites find suitable food, grow and produce their young. These migrate and penetrate into the skin as did their parents. Thus the disease spreads slowly as the parasites increase. Finally, as they become more numerous, they cover the invaded skin with a thin layer of scabs. As the disease advances the little scabs not only run together, forming one mass, but they become thicker, whitened, and hard. Later they run together over the nostrils, lips, face, cheeks, forehead, eyes and ears, and form a dry, hard, thick, scabby mass. By repeated rubbings this scab breaks up, and the skin tears, cracks, and bleeds. Later the wounds heal and scars are formed. When the scabs cover the eyelids the latter close up and the animal becomes practically blind, being unable to find its way or to see food. The insects are to be found in the moist layer underlying the scabs.

Source of contagion.—The insects which cause the disease have been derived from other sheep with which the recently infected flock may have come in contact, or which may have left a few parasites on some brush or stick, or in some trough with which the uninfected flock came in contact. The methods of infection are various, but he who understands that these parasites always come from some where else, and always from some infected flock, will soon learn what to do to prevent his sheep from becoming infected. The variety of *Sarcoptes* parasitic on sheep is similar to the variety living on goats, and it has been experimentally proven that each variety may be transferred and will live on either animal. Some of the varieties living on other animals may be transferred to sheep, but they do not thrive. It is not at all probable, therefore, that sheep are infected from other animals than sheep.

Diagnosis.—Head scab can not well be confounded with any other variety of scab. The seat of the disease and the presence of the parasite, which is scarcely visible to the unaided eye, are sufficient to definitely diagnose the malady.

Prognosis.—This variety of scab is one of the most amenable to treatment. Being mostly on the head it is easy to reach with remedies. If treated it will prove of little loss to the flock-master, while if allowed to take its course it will continue for a long period, gradually growing worse and rendering the patient more and more unsightly and ill-favored. It can cause severe inflammation of the eyes and ears. It can hinder the fattening of the animal and cause extensive alterations of the tissues of the skin. By affecting the health of the sheep, it will not only decrease its weight but materially lessen the amount of wool produced.

Medical treatment.—The worst feature of treating the disease is, that treatment for a complete eradication seems to be extreme and out of proportion to the end to be attained. Curative treatment always yields good results when the application is rational. In the early stages of the disease, when the scabs are just formed, simple applications of scab dips or ointment are all that are needed; in cases of longer standing it will be found necessary to first soften and loosen the scabs with some kind of grease or oil, and then to remove them with some alkaline solution or soap. The thin oils (sweet oil) that penetrate are the best. The wool adjacent to the scabs should be cut away so as to allow the remedies to get at the newly affected portions.

COMMON SCAB.

PSOROPTES COMMUNIS, Fürst., var. OVIS.

Plates VII and VIII.

Common scab is caused by an insect known as the scab-mite or itch insect—*Psoroptes communis*, Fürst., var. *ovis*. This insect is much larger than the *Sarcoptes*, which causes head scab, being visible to the unaided eye.

Disease.—Of all the diseases of sheep in this country, scab is the most feared by the flockmaster. So insidious is its attack, so rapid its course, so destructive its effects, and so difficult is it to exterminate that it has justly earned the distinction of being more injurious than any other disease caused by external parasites. Scab alone, of the parasitic diseases, has become the subject of legislation in most countries, and yet, if proper precautions were taken and a rational treatment followed, this disease could soon be completely eradicated.

Early symtoms.—Attention to the disease is first attracted by the infected sheep scratching, biting, and rubbing themselves. The coats of the animals look rough, taggy, and felted. The itching is always most violent when the sheep have been heated by driving or warming in a stable.

Pathology.—By separating the wool and examining a recently infected spot, there can be seen some minute elevations, which differ from the surrounding skin in being slightly whiter or yellower, and which have been produced by the bites of the pests. The insects themselves can be found among the hairs at but little distance from the bites. As time passes and the insects multiply in numbers these elevations become more and more numerous, and closer and closer together, until they finally unite over a considerable extent. From the summit of each elevation or papule, a watery, serous fluid exudes and accumulates, which transforms them into vesicles and pustules, and which in drying cover them over with a thin crust. In a few days the whole surface is covered with a yellowish, greasy, scaly layer, under which the parasites are hidden. As the disease proceeds this layer gradually increases in thickness by an increase of the serous exudate, and in circumference by the extension of inflammation produced by the ever-multiplying parasites which live beneath it, forming scaly crusts. These crusts, in being torn out, mainly by the rubbing with which the sheep endeavors to allay its intense itching, carry with them the tags of the wool, the loss of which is an early symptom of the disease. At a later period the crusts are replaced by another set of thicker, firmer, adherent scabs, which are still further enlarged by the outward migration of the parasites. As they abandon the center of the scabs these are again replaced by a peeling off of the external layers of the skin, which gradually heals, while the disease slowly progresses at the outside. The complete cure is very slow, and the skin remains thick and folded for a long time. In sheared sheep the skin becomes covered by a thick, dry crust, like parchment, while beneath it remains much swollen.

Late symptoms and diagnosis.—The fleece of scabby sheep presents a characteristic rough look. In places the wool is stuck together in masses; in others it fails, while in others, which are apparently sound, it can be easily plucked off. The rubbing and scratching indulged in by the sheep not only tend to tear away the wool but increase the irritation of the skin, which may become intensely inflamed and swollen and finally end in a superficial death of the part. Unlike *Sarcoptes*, the *Psoroptes* seeks the longest, thickest wool. It begins its attack along the back and extends to the neck, flanks, and rump. The *Psoroptes* are rarely found in the region of the chest and abdomen. They are collected in masses on circumscribed surfaces. The scabs they produce constantly increase at their edges, and their number depends on the number of places invaded. Owing to the closeness in which sheep congregate and to their violent scratching the parasites become very generally scattered and finally the scabs may run together.

While few of the parasites are present in the older diseased parts, at the edges of the scabs they can be found in swarms. They look like little white points with a brownish extremity. If picked up by the point of a knife or a sharp stick and placed on the hand they will be seen to move. The six-legged young, the eight-legged adults, the sexes, couples joined together, and the eggs of this interesting insect can easily be identified by the aid of a low-power magnifying glass.

Prognosis.—The disease is favored in its advance by the seasons in which the wool grows longest, and in which the sheep are brought into closer contact in sheds. Autumn and winter are the most favorable for its spread and rapid advancement. In summer, and especially after shearing in spring, the disease makes little, if any, headway until the wool has grown to a considerable length. Age, temperament, state of health, energy, and race of the animals, the length, fineness and abundance of fleece, and the hygienic surroundings have much influence on the advance, progress, and termination of the disease. The young, the weak, the closely in-bred, the long coarse-wooled sheep, and those subjected to bad climate, to unhealthy localities, to poorly constructed, illy-ventilated sheds, are all more subject to the rapid advances of the disease. On the other hand, healthy, well-fed, well-housed sheep may withstand the ravages of the disease for months.

When left to itself scab causes severe disturbances of the functions of the skin, and on account of the intense itching brings on fatigue, through loss of rest and sleep. Marasmus and cachexia preceding, death may come to weak, ill-nourished subjects in two or three months.

The mortality due to scab varies much, depending on the season, general health of the flock, food, shelter, and a variety of other factors. It is most disastrous in autumn and winter among sheep poorly fed and housed, and of weak constitutions. Many other diseases may intervene and carry off the weakened animals. The death-rate is not the only occasion of loss, for whether the shepherd keeps his flocks for mutton or wool he will find a loss in either, depending much on the severity of the disease. Ewes weakened by the disease will remain infertile, abort or produce but weak and feeble lambs, which will either die or scarcely be worth the rearing. To this loss must be added the decreased value of the wool obtained from the first shearing after a recovery from the attack, due to the mixing of the ends of the old wool with those of the new, known as the double-ended wool. This mixture lessens the value for manufacturing purposes.

Source of contagion.—Remembering that common scab is caused by insects which the infested flocks are continually spreading broadcast by dropping tags of wool by the wayside, by leaving them attached to brush, by rubbing posts and fences, it is easy to realize that there are many methods of transmitting the malady. Experiments with these insects have shown that they can live at a moderate temperature on a piece of scab from ten to twenty days; that they may live after being subjected to intense cold; that they die more rapidly when they are in contact with animal matter at freezing temperature, and that they die soon if they remain exposed to alternating high and low temperatures. These experiments show that the *Psoroptes* can live about the sheep sheds, yards, corrals and fences during twelve or fifteen days, at least, after they have separated from the sheep.

Although this species of parasite is but a variety of the *Psoroptes communis*, of which the *Psoroptes* parasitic on horses is another variety, the latter has not yet been made to grow on sheep experimentally. It is not probable that either of the varieties parasitic on cattle or rabbits would thrive on sheep. That is to say, so far as is now known, sheep are infected with common scab from other sheep, and can not acquire it from other animals. On the other hand, the ovine variety of *Psoroptes* has not yet been found to grow on other species of our domestic animals.

Differential diagnosis.—*Psoroptic*, or common scab, is different from *Sarcoptic* or head scab, in that the former chooses to live where the fleece is longest, and the latter where there is little or none at all—the one on the back and sides, the other on the head and occasionally on the nether parts; the one is almost invisible to the naked eye, and the other is plainly seen, though small. The itch due to other parasites, such as sheep ticks and sheep-lice, can be easily separated, because they are large and can be found in the wool. Sheep are sometimes subject to an inflammation of the sebaceous glands. In this, however, there will be no parasites of any kind found.

Prognosis.—Scab, as has been said before, is one of the most dreaded diseases of sheep. For the flockmaster who has but few sheep, say fifty or a hundred, the task of treatment and eradication of the scab from the flock is no easy affair; but for him who owns from five to twenty thousand the difficulties to be met are enormously increased. Though the disease may be easily treated as far as a single sheep is concerned, still the treatment would only be palliative and would not assure the flockmaster that the disease would not break out again. Treatment, therefore, of a flock in which scab has appeared must be applied to every individual exposed and to the corral and sheep-pens in which they have been lodged, and is not only a serious time-consuming affair, but a most expensive one.

Treatment.—This is of two kinds—preventive and curative. The preventive treatment is undertaken before, during, and after the curative. Indeed, if the flockmaster exercises proper care his flocks will never require the curative treatment, for the disease always comes from transference of the insect.

Preventive.—An infected flock should be quarantined so that it shall not transmit the disease to other flocks, and should be kept from public highways where other flocks may pass, until it can be thoroughly cleansed and cured. The sheds, yards, and corrals where they have been kept should also be cleansed, so that they may not transmit the disease. After treatment begins the sheep should be transferred to a temporary uninfected yard, so that the old yard may be thoroughly disinfected by carting away the soil to some safe spot, by washing all the wood-work as high as a man's head with a solution of boiling lye,* and afterwards covering it with a coat of whitewash. All old pelts which could harbor the insects should be burned. Every possible secreting place for the insects should be overhauled. After a thorough cleansing the yards should be left vacant for three weeks.

After the dipping the sheep should be driven into fresh, clean yards, and not into such as have not been sufficiently cleansed. If possible they should be kept from infected pasture ranges for three weeks, by which time all parasites which may have dropped from them may be considered as dead. Sheep which have been dipped in any of the tobacco preparations can be, so it is stated, driven on the ranges immediately after dipping with impunity, as the tobacco odor keeps the insects away. Any sheep which may have died on the range should either be buried deeply or burned. In dipping extreme care should be taken by all who handle sheep not to transfer the pests from animal to animal.

Medicinal.—There are two methods of treating sheep for scab. The one of rubbing poisonous ointments and oils into the fleece by the hand is the oldest and least used. It is slow, tedious, and unreliable, and has been superseded by better methods.

The other consists in immersing sheep in watery mixtures which will kill the parasites. This method being cheaper, quicker, and more effective, is the one in general use in this country. The formulæ used and the methods of applying them vary in different portions of the country according to the needs of the sheep-owner.

In the East, tubs large enough to hold sufficient of the dip to completely immerse the sheep, and kettles or cauldrons of a capacity to heat the required amount are used, but in the range country of the West, where thousands of sheep are to be treated, especially made dipping pens and tanks through which the sheep may be driven, and large boilers made for the purpose, are used. Each method is adapted to the needs of the respective localities. Although some dips are fairly effective when applied to sheep with their fleeces on, the dipping should, as a rule, be preceded by shearing. This rule should be violated only on account of season. If any of the flock are infected all should be subjected to treatment, otherwise the disease will be carried along and break out from time to time. The shearing should take place in a shed where all the wool can be cared for, and either poisoned or destroyed, or so safely stored that it could by no possibility scatter the insects. Any treatment undertaken without being preceded by shearing can at best be considered as palliative.

The object of the treatment is to kill the parasites and their eggs. The parasites are killed by the direct application of a poisonous dip. The eggs have a thick shell which often resists the effect of the poison,

---

. * Use 1 part of potash to 200 of water.

and the young parasite emerges in due time. They are then subjected to a second dip some six or ten days afterwards, at which time it is presumable that all the eggs have hatched and none of the young have become adult. If the second dip is delayed much longer than twelve or fourteen days, some of the newly hatched larvæ may have become adult and laid eggs, which may in turn hatch larvæ, and become new centers of infection.

All dips, to be effective, contain some ingredients which are poisonous to the parasite. This poisonous element may also, if used in too concentrated a solution, be poisonous to the sheep, but this is to be avoided by using the dips in the exact proportions of the formula given and maintaining these proportions throughout the treatment. In addition to the poisonous element, a dip may contain other elements, as water to dissolve and to dilute the poison ; also such a substance as alum or soda to combine with the poison, as arsenic, to make it more soluble ; or it may contain an alkali, as soda or potash, to soften the scabs when applied ; or it may contain substances which are empirically added because they have been experimentally proven of service.

The chief poisons used in the dip are tobacco, arsenic, and carbolic acid. Of these, tobacco is the favorite, because its use has not been followed by the fatality that has in times past followed the use of arsenic. Carbolic acid is too expensive to be used in large quantities, but is an excellent ingredient when only a few sheep are to be dipped.

The addition of tar to these dips is excellent, as the tar water is not only good for the wounds but serves an excellent purpose of driving away the flies.

The quantity of dip required for each sheep is variously estimated at from 1 quart to 1 gallon. For small numbers of sheep, say fifty or one hundred, the larger amount is necessary, but for large flocks, 1 quart for shorn and 2 for unshorn sheep may be allowed. It is always best to have more of the ingredients on hand than is necessary, so that they may not be used up before the dipping is finished and thus delay the business. To make the dip more effective the solution should be administered quite hot. The most desirable temperature is from 100° to 110° Fah., which is a comfortable one for the sheep, whose internal temperature is about 103°. The warmth enables the dip to penetrate the oily wool better, makes the parasites livelier, and proves far more efficient.

Instead of treating the scab by one application some authorities advise the use of a preliminary dip of alkaline water to soften the scabs, or of oil or glycerine well rubbed in for the same purpose. This is to be followed in two or three days by a poisonous dip. Nearly all advise that the scabs should be rubbed with a stiff brush while the sheep is being dipped.

The Australian or Rutherford dip, which has been very successful in the hands of large flockmasters, is as follows : Take of tobacco and flowers of sulphur 1 pound each, to every 4 gallons of water to be used. The tobacco should be steeped in a portion of the water two or three successive times so as to extract all of the juice. The leaves-or stems may be used; of the latter three times the weight is required as is needed of the former; a press or wringer is convenient to squeeze out all of the liquor from them. The sulphur should be mixed with some of the tobacco water and stirred until it is of creamy consistency. These ingredients should be added to the required amount of water. During the dipping this mixture should be constantly stirred and a little fresh water added from time to time to replace that lost by evaporation.

This dip, to be more effective, should be heated to between 100° and 110° Fah. in summer, and 110° and 120° Fah. in winter, never being allowed to fall under or exceed these limits. The sheep should remain immersed in it from sixty to ninety seconds, and the head should be completely immersed at least once.

When sheep with heavy fleeces are dipped it will be found necessary to separate the fleece with the hands, that the fluid may permeate better. Eight or ten days after the first dipping the treatment should be repeated. Sometimes a third and more rarely a fourth dipping is necessary. When the last two are required it is most often due to carelessness in preparation, or a failure in the strength of the first dips. Sometimes it may be necessitated by the rain having washed off the first solution soon after dipping.

. Australian sulphur and lime dip: Take of flowers of sulphur 100 pounds, of quicklime 150 pounds, water 100 gallons. Mix and stir while boiling for ten minutes, until the mixture assumes a bright red color, then add 3 gallons of water. Hold the sheep in the mixture until the scabs are thoroughly soaked. Immerse the head at least once. Use the dip at 100 to 110° Fah. Dip twice at an interval of two weeks.

In the American Merino, 1887, Stephen Powers gives an excellent description of sheep-dipping on a large scale, and the following recipes in use in various sections of the United States:

Texas and New Mexico: Thirty pounds of tobacco, 7 pounds of sulphur, 3 pounds of concentrated lye, dissolved in 100 gallons of water.

Nevada: sulphur, 10 pounds; lime, 20 pounds; water, 60 gallons.

California: Sulphur, 4 pounds; lime, 1 pound; water, enough to make 4 gallons.

Kansas: Sulphur, 22 pounds; lime, 7 pounds; water, 100 gallons.

Sulphur and lime is probably the cheapest recipe, but the lime is apt to injure the staple; still this recipe appears to prevail over all others in the scab-infested regions. Probably tobacco and sulphur form the best combination known for the treatment of scab. To every hundred gallons of water there should be used 35 pounds of good strong tobacco (if stems or other inferior parts are used there should be more), and 10 pounds of flowers of sulphur. This preparation used at a temperature of 120° Fah., will kill all acari ticks and lice, and leave the wool in a healthy condition. To insure thorough work apply a second time in ten days or two weeks.

Walz's dip, one of the oldest recipes, is as follows : Take of fresh slaked lime 4 parts, carbonate of potash 5 parts, mix and boil in barn-yard water; add animal oil 6 parts, tar 3 parts; dilute with barn-yard lees 200 parts, water 800 parts. To-day this recipe is more curious than useful; one of the ingredients is not on the American market, and another is offensive. Veterinarian Clok, in his Diseases of Sheep, 1861, reports this mixture as being too weak for old cases, and recommends the following modification: Take of freshly-burned slaked lime 6 pounds, add potash 6 pounds, and water 10 quarts; boil an hour, stir-ring occasionally. Add pine-oil 8 pounds, and tar 2 quarts, stirring the mixture thoroughly. Make an infusion of 20 pounds tobacco in 130 quarts of water; add the lye already made and stir. This quantity suf-fices for one hundred sheep. Apply by immersing the sheep, separat-ing the wool and breaking the scabs. Repeat in eight or ten days.

Law's recipe (Farmers' Veterinary Adviser) is a very good one: Take of tobacco 16 pounds, oil of tar 3 pints, soda ash 20 pounds, soft soap 4 pounds, water 50 gallons. This quantity suffices for fifty sheep. The tobacco should be steeped; afterward the other ingredients should be added at 70° Fah.

Zundel's dip is available, but Dr. Kaiser (*Kuhrverfahren bei der Schaf-räude*, 1883) reports that it is too weak in cases of long standing. It is said to leave the wool clear and white after using. For every one hun-dred sheep take crude carbolic acid 3 pounds, caustic lime 2 pounds, pot-ash 6 pounds, black soap 6 pounds, and water 70 gallons; mix and boil. Dr. Kaiser has obtained excellent results from a modification of this recipe. Take of tobacco 5 kilograms (13½ pounds), infuse it in 250 liters or 66 gallons water; dissolve in it 3 kilograms (8 pounds) soda, add 1½ kilograms (4 pounds) freshly burned and slaked lime. Dilute 3 kilo-grams (8 pounds) black soap (soft soap will do) with hot tobacco broth and add it to the rest; then add 1½ kilograms (4 pounds) crude carbolic acid, which contains at least 50 per cent. of the pure acid. Mix. This quantity is sufficient for one hundred sheep.

Gerlach's dips, which are administered at two different times for a single treatment, are cheap, but owing to the time and labor to be ex-pended in performing an operation twice which in other instances is done but once, is much more costly than those which require but one application.

Take of potash 2 parts, burnt lime 1 part, water 50 parts. Mix. Use this dip for softening scabs. Follow it in two or three days by the fol-lowing: Make an infusion of tobacco 34 pounds in 66 gallons water, by slowly steeping the tobacco in a portion of water and finally adding it to the remainder. Repeat the dip in from six to ten days.

Roloff's dip, for a mixture sufficient for one hundred sheep, is : Take 7.5 kilograms (20 pounds) of tobacco, steep it with 250 liters (66 gal-lons) water for half an hour, heat it to 30° R. (95° Fah.) and add 1 kilogram (2½ pounds) each of pure carbolic acid and of potash.

A cheaper and fully as effective dip of similar formula is the following: Mix an infusion of 15 pounds tobacco with 1 kilogram (2½ pounds) carbolic acid and 5 kilograms (13½ pounds) wood tar, pour it into 250 pounds (66 gallons) water at 40° R. (125° Fah.), in which 3 pounds soda has been dissolved. Use it at a temperature of 80° or 90° Fah., and repeat in six or seven days.

There are three arsenical dips favored by European authorities, the last of which might be used while exercising proper care and precaution. The solution should always be kept as dilute as the formula calls for. The sheep should not be allowed to drain on the grass, but should be kept up until nearly dry, and the laborers who dip should grease their arms with linseed-oil before beginning work.

Tessier's dip, the oldest, was proposed in 1810. To make a mixture for one hundred sheep, take arsenious acid 3 pounds, sulphate of iron 20 pounds, and water 190 pounds; boil.

Tessier's dip causes a discoloration of the wool, which can be removed by washing with soap, but it is on this account more or less objectionable, and to overcome this objection Clément has modified it as follows: Arsenious acid, 1 part; sulphate of zinc, 5 parts; water, 100 parts. The water is put over the fire, the medicinal substances added, and it is allowed to boil for eight or ten minutes. After the temperature has fallen to the proper degree it is ready for use. The sheep is entirely submerged in the liquid, with the exception of the head, the udder of ewes suckling their young having been previously covered with some fatty substance to prevent the action of the astringent on the skin and on the secretion of milk.

Matthews dip: Take arsenious acid 1 part, alum 10, and water 100 parts.

Scheurle and Kehm's dip: Take arsenic 1 part, alum 12, and water 200 parts. This latter is weaker than the former, and therefore safer. Moreover, it is claimed to be as effective.

The sheep dips that are put upon the market are objectionable for three reasons: First, their formulas are not given; second, the preparation may be valueless, or if not valueless of insufficient strength; and third, the preparation may cost more than it is actually worth. Should the dip be put up by reliable houses, and have their formulas printed on the outside of the package, the prepared article might prove better compounded and absolutely cheaper than the flockmaster could prepare it.

Police sanitation.—It is not sufficient that the flockmaster thoroughly cures his flock, disinfects his sheds and quarantines his place. Another duty awaits him, and that is a public one. All flock-owners should unite and assist the State in improving and carrying out its sanitary laws. Nearly all States have laws regarding the suppressing and quarantining of sheep infected with scab, but they seem to lack in stringency and are therefore inadequate. Laws compelling strict quarantine and public supervision of the treatment, at the expense of the owner, even though it is undertaken by the State, are demanded for the complete eradication of the disease. With such laws no one could long harbor on his premises a disease which constantly threatens the flocks of his neighbors with destruction and their owners with financial ruin.

FOOT SCAB.

CHORIOPTES COMMUNIS, Verheyen, var. OVIS.

This variety of scab, which is due to *Chorioptes communis*, Verheyen, var. *ovis*, is of rare occurrence. It has been noticed and studied in Germany by Zürn.

The seat of this disease is in the feet and limbs. The disease progresses very slowly from the feet, and little by little invades the upper parts of the limbs and adjoining parts. It is not readily communicable to other sheep and spreads slowly.

In the beginning this variety of scab is characterized by the reddening, followed by an abundant scaling of the skin, and later by yellowish white crusts. The animals stamp, scratch, and bite the parts, showing an intense itching. As the disease progresses the crusts become thicker; cracks may form in the folds of the pastern and the limbs become quite unsightly. The parasites swarm beneath the crusts, and when found form a certain symptom of the character of the disease.

Foot-scab is not a serious malady, as it readily yields to treatment and is of slow extension. Any of the remedies proposed for the treatment of common scab may be used with good effect.

23038 A P——5

PSOROPTES COMMUNIS Fürst., var. OVIS.

PLATE VII.

Fig. 1. Adult male, dorsal view : *a*, head ; *b b*, legs; *c c*, suckers.
Fig. 2. Adult male, ventral view.
Fig. 3. Adult female, dorsal view.
Fig. 4. Adult female, ventral view.

NOTE.—Figs 1 to 4, Plate VII. and Figs. 1 to 3, Plate VIII, are equally magnified.

PLATE VII

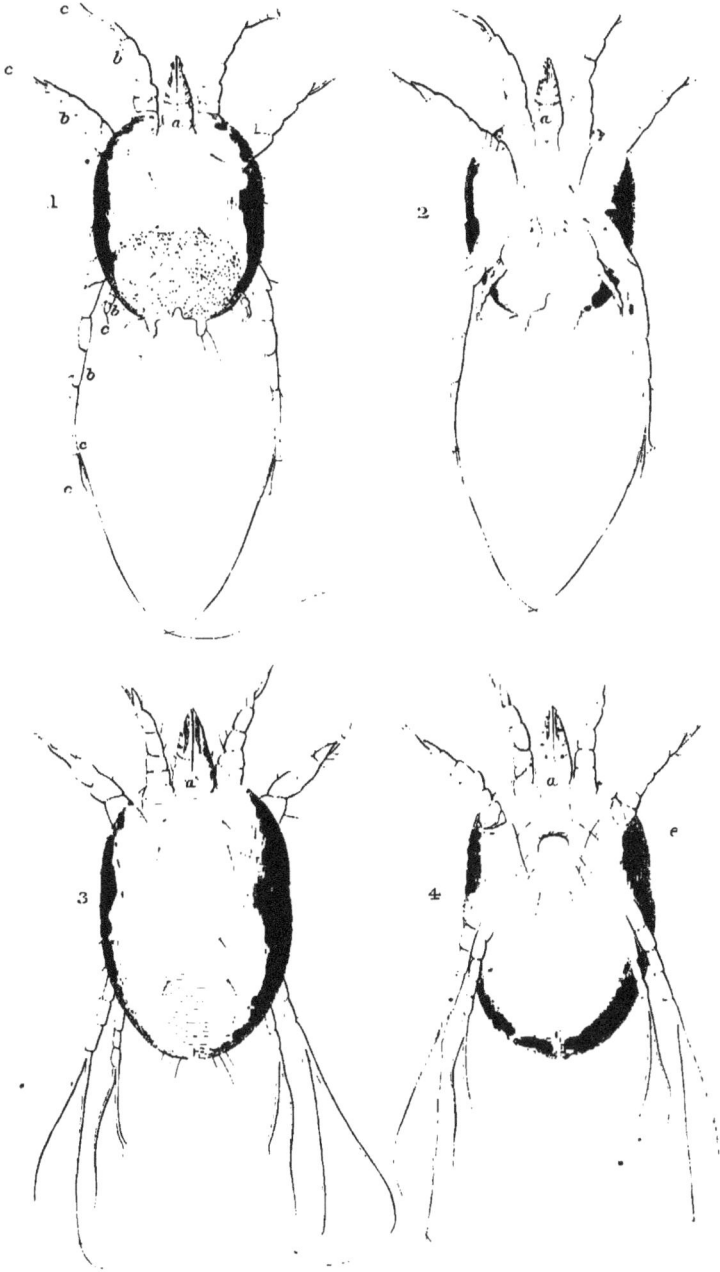

PSOROPTES COMMUNIS, *Var* OVIS,
(The Common Scab-Mite.)

PSOROPTES COMMUNIS Fürst., var. OVIS.

PLATE VIII.

Fig. 1. Young female before moulting for the last time.

Fig. 2. Egg drawn from a specimen which was inside an adult female.

Fig. 3. Young six-footed larva.

Fig. 4. *a*, open, and *b*, closed sucker of *Psoroptes* from ears of rabbit; *c*, the rod which connects the membrane on the end with the muscles which close the sucker.

Fig. 5. Two views of the mandibles. The lateral spurs, *a, a*, point outward (Megnin).

Fig. 6. Head and anterior limb enlarged; ventral view: *a*, mandibles; *b*, antennæ; *c*, maxillæ; *d*, membrane joining the antennæ; *e, e, e*, joints of the limb; *f*, the claw; *g*, the ambulacrum or sucker.

Fig. 7. Male and female of *Psoroptes communis* var. *equi* (Megnin).

PLATE VIII

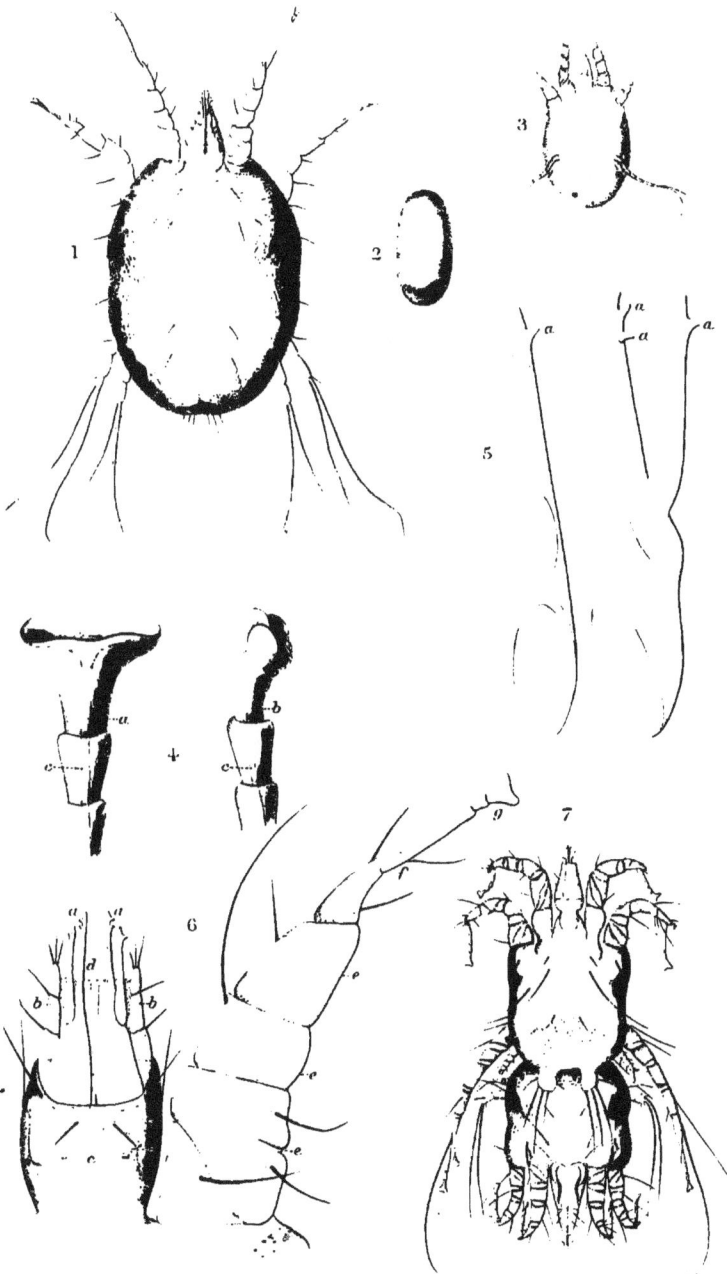

PSOROPTES COMMUNIS, *Var* OVIS.
(The Common Scab-Mite.)

THE PENTASTOMA.

LINGUATULA TÆNIOIDES, Rud.

See Plate XVII, figs. 1-6.

Linguatula tænioides is also known under the names *Linguatula rhi-naria*, Pilger, and *Pentastoma tænioides*, Rud.

In describing this parasite, Neumann's excellent description in *Traité des Maladies Parasitaires*, page 491, has been taken as a guide. The species is probably present in this country, as in two instances the larval form known as *Pentastoma denticulatum* has been found. The larvæ were found by Dr. F. L. Kilborne, of this Bureau, in a rabbit, some time in the summer of 1887, and once by myself in another rabbit in 1888.

This curious parasite is classed among the Arachnids, being more closely related to the spider and mite family than to any other. The order *Linguatulidæ*, to which they belong, is thus defined:

Endoparasitic arachnids with elongate, vermiform, annulate body. Mouth wanting jaws in the adult state, and surrounded by two pairs of hooks representing rudimentary feet. No heart. Respiration cutaneous. In it are two genera: (1) *Linguatula*, Fröhlich, which has the body depressed with dorsal face rounded and with the borders crenulated. The cavity of the body forms diverticula in the lateral parts of the rings. (2) *Pentastoma*, Rud., which has a cylindrical body and the cavity of the body continuous. It is not represented in our domestic animals, unless it be a larval form found once in the peritoneal cavity of a dog.

Description.—The species met with in the domesticated animals has the following characters:

Body white, lanceolate, very elongate, worm-like, depressed dorso-ventrally with the ventral face nearly plane and the dorsal face convex. Anterior extremity rounded, large; posterior extremity attenuated. Cephalo-thorax short, solidified in all its width to the abdomen, from which it is scarcely distinct, and which forms by far the larger part of the body. Integument showing about ninety rings, larger in their middle; these make the borders of the body crenulated. Hooks acute, recurved, twice-jointed, the basal joint attenuated in its deep part. These hooks, each retractile in a pocket, are moved by muscular bundles which act in opposite directions. Mouth sub-quadrangular, rounded at the angles; digestive tube simple, rectilinear; anus terminal. Male, white: length from 18 to 20^{mm}; width, in front, 3^{mm}; behind, 0.5^{mm}, provided with saccular testicles which fill the body cavity even to the anterior fourth. Female, whitish grey, often rendered brown by the eggs along the median line where the integument is thin and semi-transparent; length from 8 to 10^{cm}; width, in front, from 8 to 16^{mm}, and behind, 2^{mm}. Eggs ovoid; length, 0.09^{mm}; width, 0.07^{mm}.

Life history.—The female *Linguatula tænioides* lays its eggs in the nasal cavities of the dog. These are scattered on the ground and grass where they lie until eaten by some herbivorous animal. The shell is then dissolved from around the embryo, and it bores through the walls of the stomach or intestine into the mesenteric gland, liver, or lung, where it encysts itself. In its first stage of active migration the larva resembles the Acari (Plate XVII, Fig. 4). It has an ovoid body, flattened on the ventral face, rounded on the dorsal. Its posterior extremity is narrowed and dentate. It is furnished with two pairs of articulated, two-clawed feet, and at its anterior end by a perforating apparatus formed of a median stylet and two re-curved hooks. Its length is 0.13^{mm}; its width 0.06^{mm}.

Having arrived at the mesenteric glands, the liver, or the lungs, as the case may be, the embryo loses its feet and is transformed into an immovable pupa (Fig. 5), without segments, hooks, or hairs, measuring 0.250 to 0.300^{mm} long, and 0.180^{mm} in width.

It emerges from this cyst transformed into another larva by a series of successive moults (see Fig. 6). The body is elongate, larger forward, and is divided into eighty to ninety rings bordered behind by a series of fine spines. The digestive tube is large, the mouth is elliptical, and surrounded with four characteristic hooks and with accessory hooks. The larva is agamic, its genital organs being rudimentary and represented only by a little granular mass in the posterior part of the body. Towards the sixth or seventh month the larva is completely developed, measures 6 to 8^{mm} long, and is in the stage called *Linguatula denticulata*.

These larvæ having escaped from the cyst, fall into the serous cavities and remain there for some time. They eventually escape, but the precise method is unknown. Next they are seen in the nasal cavities of dogs. Exceptionally, so it is said, they are found in the nasal cavities of sheep and cattle, into which they have wandered. These larvæ can acquire their full development only in the respiratory passages. Once installed in the nasal cavities they develop into egg-bearing adults. The males wander and can be found at various points of the cavities, but the females are more sedentary, and are never found in the ethmoidal cavities. After the death of the host they may travel into the pharynx and larynx. They exceptionally introduce themselves into the frontal sinuses. They are generally found at the bottom of the nasal chamber.

As the adult stage is not usually found in sheep, and as its occurrence is problematical in this country, the disease it causes will not be considered in this volume.

Disease.—The young state, *Linguatula denticulata*, found in cysts within the glands, etc., are said to be particularly frequent in sheep in Europe. Sheep in which the parasite affects the mesenteric glands are generally less fat; their flesh is paler, and they are apparently predisposed to anæmia. These glands show no evidences of the parasite at first, but later they grow browner, smaller, and are crossed by galleries filled with larvae. These cavities are separate at first, but finally communicate; the substance of the gland is destroyed and transformed into a brown tumor, in the middle of which are the *Linguatulæ*. From these the parasites often escape through openings with irregular borders; at other times the surface is covered by dark, irregular spots, fibrinous deposits, and false membranes, which indicate a recent departure or a de-

struction of the embryos. At last the tissue is found thickened, indu-
rated, and offers here and there tuberculoid grains formed from the old
nests of *Linguatulæ*. These altered glands are destroyed and are of no
further use in nutrition. The larvæ create further troubles in their migrations, but so little is known about these parasites that little can be
said.

There is no treatment. Prevention is also difficult. If it should be
learned that we have these parasites here in considerable numbers the
best remedy would be to remove the dogs, which are certainly the cause
of large numbers being scattered in Europe.

IMMATURE TAPE-WORMS—BLADDER-WORMS.

Plates IX, X, and XI.

Besides the adult tape-worms found in the intestines of sheep, there have been four other species described which infest various portions of these animals in their immature stages. These species are *Tænia marginata*, Batsch, *T. coenurus*, Küch., *T. echinococcus*, v. Siebold, and *T. tenella*, Cobbold. The forms found in sheep were first described as *Cysticerci*, and have since been known as *Cysticercus tenuicollis*, *Coenurus cerebralis*, *Echinococcus* and *Cysticercus ovis*, respectively.

All these species resemble each other in their anatomical structure, their growth, and their life history. They differ in minute structure, in invading different portions of the sheep, and in the effects they produce on the animal.

Tænia marginata is more common in the United States, and *T. coenurus* next. Neither of the other two species have been found in sheep in this country.

TÆNIA MARGINATA, Batsch.

Plate IX.

Tænia marginata occurs in sheep as a little semi-transparent bladder filled with liquid, varying from a very minute size to a bag an inch or more in diameter, but usually having a diameter of a half or three-quarters of an inch.

Occurrence.—This *tænia*, in its *cysticercal* stage, is usually found between the layers of the serous membrane which form the omentum, or "caul" of the abdominal cavity. It may be found in the liver, especially within a week or two after the infection of the sheep by it. A very common place to find it is between the folds of serous membrane which suspends the intestine in the pelvic cavity. When one of these little fluid-sacs are found it may be cut out, with the surrounding tissue for examination. Afterwards great care should be taken in dissecting the serous tissue, which forms an outer sac, so that the *cysticercus* within shall not be cut, its contents allowed to escape and its walls to collapse.

Description of cystic stage.—The smaller bladders are apparently composed of a semi translucent whitish membrane, at one end of which may be seen a whitish thickening. As the bladders grow the walls become slightly thicker and the spot at the end becomes much larger and projects in the form of a knob.

72

When the animal is placed in a saucer of lukewarm water immediately after its removal from the slaughtered sheep and examined, it can, by the aid of a low-power lens, be seen to possess considerable peristaltic movement. This movement is produced by bundles of muscles lying at right angles to each other, which may be seen appearing as a faint striation on the surface.

The little knob end of the bag is its essential part, and contains what is to develop into the future *tænia*. Sometimes the animal will extend this knob into a cone, and finally thrust out of its center the very tip of the cone. By careful handling this so-called head end may be squeezed out by the fingers. The tip, when examined by a magnifying glass, can be seen to possess four cup-like spots, with a little glittering circlet of hooks between them at the very apex of the cone.

Life history.—When these cysts have attained their hooks and cups in a well developed condition, they are ready for transplanting into another animal or host. The *cysticercus* completes its development in about eight weeks. It may live a long time after this, and its cyst enlarge, but the modifications it may undergo are unessential. The host within which the cysts or *cysticerci* generally develop is the dog. They may, however, also develop in other carnivora, such as the wolf or coyote.

Their emigration is a passive one. They remain encysted where they are found until the sheep is slaughtered and the dogs eat the offal, or until the sheep is killed by a dog or wolf and its liver is torn from its place and devoured, together with any of the *cysticerci* which may be attached. Having gotten into the intestines of the host the parasite completes its development, becomes adult, and finally produces young, which pass from the host along with the ejecta of the intestines.

The young at this stage are egg-like. They are very small and hardly visible to the naked eye. When viewed with a glass they are seen to be a minute, jelly-like mass, furnished with six hooks and surrounded by at least three membranes. The outer is thin and filled with fluid; the inner two more closely surround the embryo, and confine between them an oily material which serves to protect it when exposed to the atmosphere.

After passing to the ground these embryos in some way, possibly by adhering to food or by floating in drinking water, make their way into the sheep. When they arrive in the abomasum, or fourth stomach, it is supposed that the gastric juice digests the membranes surrounding the embryos and they then begin their active wandering. At this stage they penetrate the walls of the fourth stomach and make their way between the walls of serous membranes to the place where they finally find lodgment. This active migration must often be converted into a passive one after the embryo has made its way through the mucous coat of the stomach, for when the embryos have been fed in large numbers to the sheep in the course of the experiments of different investigators the liver has been found to be filled with multitudes of them. One investigator has found them in the minute branchlets of the portal vein, which conducts the blood from the stomach and intestines to the liver. Now, in order to appear in the liver and in the portal vein in such numbers the embryo must make its way into the little branches or capillaries which collect the blood at the stomach, and then be washed by the blood current into the liver. When they become lodged in the liver they again migrate actively and tunnel through the mass of the liver in all directions. The little channels are made slowly. They begin as a minute point, and become gradually larger as the parasite increases in size and changes its position. About the ninth day after the embryos have been swallowed some of the parasites may be as large as a flax-seed; most of them will be smaller. They will then be little oval water-bags, with a whitish thickening at the end.

Some of the parasites seem to pass the entire length of the intestine before they penetrate the walls; but when they reach the rectum they pass through, and, becoming lodged between the layers of serous membrane, develop there. This probably accounts for the presence of the considerable numbers which are found in the pelvis.

The destination at which these parasites arrive has its influence on their development. When they have migrated to a point between serous tissues which may be easily spread apart, and offer little resistance to the growing parasite, a certain proportion of them seem to thrive and arrive at a stage in which they may continue development by the proper exchange of hosts. If this exchange is not effected the parasite may remain unharmed for a long time while awaiting this opportunity. Should they die from any cause a slow change in the appearance of the parasite is noticed. The fluid it contains becomes milky and limy. The serous sac surrounding it becomes thicker; finally such changes have occurred that in place of a soft sac a round, hard sphere of lime covered by a thick membrane may be found. The remains of the parasite may still be seen by careful dissection between the cover and the limy deposit.

But when, on the other hand, they arrive at the liver they seem to have reached a place of destruction; for if they do not kill the invaded host within two weeks, a period long before they could continue their life in another host, the same degenerative processes seem to affect them, in their earlier stages, which destroy the older individuals in other organs of the body. This seems in part due to the pressure which the liver cells and capsule exert upon them. These statements seem substantiated by the presence of either scars or calcareous nodules in the liver and of no large *cysticerci*, except where the loose serous membrane may have permitted their growth.

These parasites have occasionally been observed in other parts of the system than those mentioned, as in the lungs, heart, and muscles.

Disease.—The harm that *Tænia marginata* does in its young stage depends much on the degree of infection. In experiments animals have been killed in from nine to twelve days after feeding. In such cases the sheep usually die of hemorrhage from the liver and peritonitis. This is caused by the perforations of the serous covering of the liver due to the parasites. These experiments show that the parasite is injurious to sheep. The presence of a few well-advanced *cysticerci* in nearly all of our sheep also shows that though the parasite may be injurious in its young stages, the sheep rarely succumbs to them in their period of invasion. After a period of about two weeks the sheep seem to have no discomfort from them.

Sheep may be invaded by the *tenia* at almost any time of the year. The winter season, when the embryos which have been scattered by the dogs become frozen, seems to be less favorable. If the dogs have access to the sheep-pens even this season will not delay the infection. Lambs and yearlings appear to be most subject. Three months' lambs are the youngest in which the *cysticerci* have been found. Experiments in attempting to produce the cysts in old sheep by feeding generally fail. Young sheep furnish the best specimens, while in five or six year old sheep the *cysticerci* seem to have degenerated into calcareous nodules.

The adult *tenia* live in dogs. Soon after the cyst is swallowed the tissue is digested from it, and the large, watery bag, which has probably already lost its fluid in the mastication of the food, disappears. The head remains, and passing from the stomach into the intestine attaches itself by its hooks and cups, which serve as suckers. Here the parasite, bathed in the intestinal fluids which nourish it, proceeds with its development. In a day or two it lengthens and begins to show cross lines, which indicate the points where the division into segments is to occur. Later the segments appear and the worm gradually matures the organs of reproduction in them. After ten or twelve weeks have passed, the parasite begins to lose its end segments, which have fully developed embryo in them, and are ready for the infection of sheep.

Description.—*Tænia marginata*, Batsch, or *Tænia cysticerci tenuicollis* Küch., is the largest of the *tenia* infesting dogs. It is usually, when mature, about a yard long. Large specimens may be stretched, when fresh, until they are 2½ yards long. The greater number of specimens are, when contracted, under a yard long. The width of the worms also depends on their degree of contraction, the more contracted specimens being the widest. The average of relaxed specimens is about two-fifths of an inch. The same specimens, when alive, might contract so much as to measure three-fourths of an inch. The width of the worm gradually widens from its head for four-fifths of its length, and then contracts slightly to the end. This species is moderately thick, measuring a little over a twenty-fifth of an inch where thickest. When first taken its appearance is white and opaque, but when kept in water it becomes partially transparent. It then resembles a whitened, pliable band of gelatine. At its small end is the so called head, which is separated from the body by scarcely any constricted neck. The head is quadrangular, is about one-twenty-fifth of an inch in diameter, and has a circlet of from thirty-two to forty hooks at its apex, and four moderate sized suckers. The specimen figured contained twenty-eight. These hooks and suckers serve to anchor the parasite to the intestinal walls. The neck is short, and at a little distance from the head shows the division into segments which become plainer and plainer towards the end of the body. The segments are at first very short and broad, but gradually lengthening become square where the animal is widest when relaxed, and about twice as long as broad at the end. The terminal segments measure about one-fourth of an inch wide and one-half inch long. There is no alimentary tract, the office of absorption of food being filled by the skin. Motion is effected by muscular bundles situated beneath the skin, which give a variety of vermicular movements. A primitive nervous organization has been described. The entire length of the worm is traversed by two canals connected in each joint by a cross system. This system is said to serve as an excretory apparatus. It has been called a vascular system. Each adult segment contains a complete hermaphroditic generative apparatus. They begin to appear in the first third of the animal and gradually mature toward the last segment. At about the middle the eggs become fertilized and develop until the end. The terminal segments when ripe separate and pass away. In this way the segments, which were once near the head, become central and finally terminal, growing adult, maturing their embryo, and passing away in turn. The embryonic shells are 0.036^{mm}, about one-seven-hundredths of an inch thick. The embryo is six-hooked. These worms mature and liberate segments in the dog in about ten or twelve weeks after the *cysticercus* has been swallowed.

Occurrence.—The marginate tape-worm is found at about the middle of the small intestine, therein differing from *Tœnia serrata*, *T. cœnurus*, and *T. cucumerina*, which live nearer the end of the small intestine. They may be found in dogs of any age. Shepherd or collies, hounds, and slaughter-house dogs are most liable to be infected. City dogs, which have but little opportunity of being infected, rarely contain the parasite.

Disease.—The harm that the parasite does the dog seems to be inconsiderable. Were it not that the *cysticercal* stage does harm to sheep, it might well pass unnoticed by the flock master. The ease with which one can rid his dogs of the parasite seems to place the destruction of the species in his hands. For *Tœnia marginata* alone the administration of an effective tænicide every two months would be sufficient; but as *T. cœnurus* develops in three or four weeks, it would be best to treat for both at once and dose the dogs more frequently.

Diagnosis.—Every sheep-owner should proceed to dose his dogs with worm medicine, whether they are known to have tape-worms or not. The diagnosis of tape-worms in dogs is difficult, those having many often showing no symptoms. Sometimes they may be suspected from the leanness of the animals or from the little white segments they pass. Such are the benefits arising from thorough medication that the time and expense given is well laid out.

Medical treatment.—The method of treatment consists in tying the dog and withholding its supper, not water, over night. Some administer a physic at this time. The special worm medicine chosen should be given on the next morning, and followed two hours later by a dose of physic. The worms, if the treatment has been effective, should be voided inside of twelve or eighteen hours. The dog should be fed sparingly for a day or two preceding the operation. The preparatory physic may be omitted. The dog may be fed at noon on milk or meat. He should not be loosened until the following day. The excreta passed should be burned or buried in some safe place. The method of administering the medicine is as follows: A man of whom the dog is not afraid should place himself in a corner and back the dog between his legs. He should then grasp the muzzle gently but firmly in one hand; with the other hand should pull out the loose cheek at the corner of the mouth. Into the pocket thus formed an assistant can put the medicine, a dessert-spoonful at a time. The lips should then be held close, and the dog will usually move his tongue sufficiently to swallow the dose. Should he refuse, his nostrils may be closed for a second or two until he gasps for breath, when the medicine will be swallowed. If any considerable quantity is to be given the operation should not be hurried, and should be persevered in with care and patience. Some dogs will eat their medicine mixed with milk or soup, while others are very fastidious.

Areca nut has proven itself the best tœnicide for dogs. The freshly ground powder is the best. The rule for measuring the dose is to allow two grains of the powder to each pound of the dog's weight. The powder is thoroughly stirred with soup or milk. If refused, another dose may · be prepared by mixing it with butter or molasses until the mass is quite soft, and administered by hand. Follow the medicine in two hours by a table-spoonful of castor-oil for a moderate sized dog. The oil can be given alone or well shaken and mixed with three times the quantity of milk.

If areca nut is not available, a dose consisting of a tea-spoonful of turpentine and two table-spoonfuls of castor-oil, thoroughly mixed with a coffee-cup full of milk, may be substituted. This dose is one for an average-sized farm dog. The final physic is not given in this case. A 2-ounce dose of castor-oil will bring away portions of the tape-worms, and sometimes the heads, without the aid of other worm medicine.

Finlay Dun recommends the following prescription: Take 20 drops of the oil of male-shieldfern, 30 of the oil of turpentine, and 60 of ether. Beat them together with one egg and give to the dog in soup.

Zürn advises the use of about 4 drams of freshly powdered areca nut for a very large sized dog, 2½ for a medium sized, and 1 dram for a very small dog. The powder is to be rubbed up with butter. Follow in two hours by a table-spoonful of castor-oil.

Dr. Hagen advises treatment of all sheep dogs each spring and fall, thereby claiming an increased immunity for the sheep from the *cysticerci*. He recommends the following: Take of the oxide of copper 80 grains, of powdered chalk and Armenian bolus 40 grains each ; of water sufficient to mix the ingredients into an adherent mass ; divide into one hundred pills ; administer one three times daily for ten days by crushing them in a piece of meat or bread and butter.

In addition to other remedies Röll prescribes the following : (Each dose is for large dogs. For smaller ones proportionately less ought to be given.) (1) Extract of male-fern and the powder of male-fern, 2 drams each. (2) A decoction of 2½ ounces of pomegranate-root bark in water, reduced to 6 fluid ounces, and add 1 dram of extract of male-fern, to be given in two doses an hour apart. (3) From one-half to 1 ounce kousso formed into pills with honey or molasses, and a little meal. (4) From 1½ to 2½ drams of Kamala stirred with honey or water, and given in two doses inside of an hour. He advises a preliminary preparation by feeding the dogs sparingly for two or three days previous on salted food, and the administration of castor-oil the evening before. The remedies proposed are to be mixed with some material to make them fairly acceptable to the patients. With the exception of kamala, which acts as a cathartic, all should be followed in two hours by castor-oil.

After any treatment the patients should be fed with some liquid diet on the first day. After this they may receive any wholesome food.

The necessity of repeating a treatment depends entirely on the efficacy of the first, and the care exercised in preventing a re-infection. If the treatment has been successful in removing the worms, heads and all, of course no further treatment will be required. If only portions have been removed, then another dosing is necessary. For *tænia marginata* another treatment need not occur under eight weeks, for the tape-worm is harmless as far as sheep are concerned up to that period, for, as far as is known, the worm will not throw off segments before that time. For *T. cœnurus* the treatment should be repeated in about two weeks.

Preventive treatment.—The great resource of the flockmaster lies in prevention. In this he has nearly absolute control over the health of his sheep, in so far as *Tænia marginata* and *T. cœnurus* are concerned. As the dogs can only get these tænia from eating viscera of the sheep, all the viscera of slaughtered or dead sheep should be withheld from them, and either buried, burned, or rendered.

Police sanitation.—Sheep-killing dogs should be destroyed. Each owner should keep his dogs at home, so that all strange dogs may be killed in order to prevent them from harming sheep and scattering the tænia eggs far and wide over the pastures and in the drinking places. Dog laws ought to be made sufficiently stringent and adequate for the protection of sheep.

TÆNIA MARGINATA, Batsch.

PLATE IX.

Fig. 1. Adult tænia, natural size: *a*, head (the constricted part following is the neck); *b*, point at which the reproductive organs begin to develop ; *c, c, c,* adult segments either separated or about to separate.

Fig. 2. Head: *a*, the crown of twenty-eight hooks; *b, b,* the suckers or bothria; *c,* the neck, which is transversely wrinkled, but not segmented.

Fig. 3. Crown of hooks enlarged : *a, a,* large hooks; *b, b,* alternating small hooks.

Fig. 4. Form of single hook.

Fig. 5. Bladder or *Cysticercus tenuicollis* (taken from the omentum or caul of a sheep), which produces the tænia when fed to a dog: *a,* the head; *b,* the bladdery portion filled with fluid.

Fig. 6. Head of Fig. 5, enlarged, shows the opening from which the true head may be evaginated.

Fig. 7. Head of the cystic worm, evaginated, shows the suckers and crown of hooks.

Fig. 8. Segment from near *d* Fig. 1 shows the branching uterus which is filled with ova.

PLATE IX

TAENIA MARGINATA,
(The Margined Tape-Worm of Dogs.)

PLATE I.

.

TÆNIA MARGINATA, Batsch.

PLATE X.

Fig. 1. Portion of a liver from a lamb which died nine days after feeding with the embryo of *Tænia marginata*, drawn from an alcoholic specimen; natural size. The surface is seen covered with short ridges, which run in all directions, and are more numerous toward the thin edge of the liver.

Fig. 2. Cross-section taken at *a a*, in Fig. 1, shows the round cross-sections of the burrows made by the young embryos; also oblique and longitudinal sections of them. The lower edge is rough and terminates in liver substance, not in the capsule.

Fig. 3. Young embryo or *cysticerci* : *a*, natural size ; *b*, ×6.

Fig. 4. Embryo in the state in which they are fed to the lamb to produce the disease. (Leuckart), ×550.

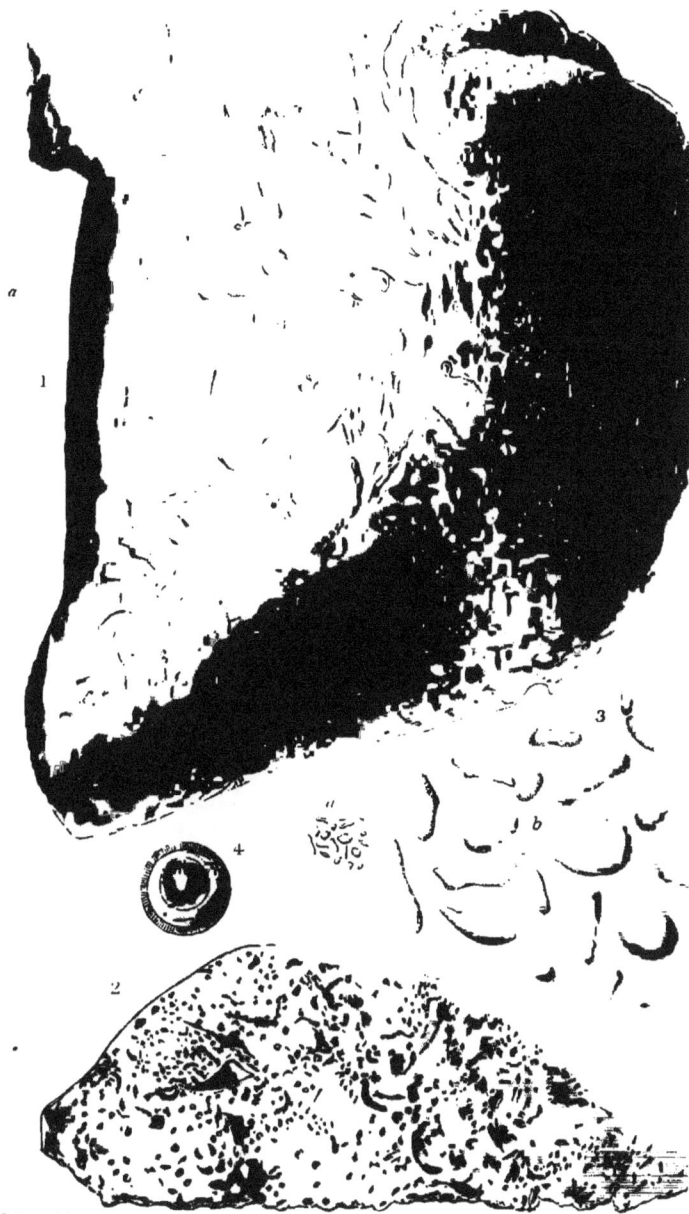

Haines, del.

LIVER OF LAMB INVADED BY YOUNG TAENIA MARGINATA.

GID OR STAGGERS.

T.ÆNIA CŒNURUS, Küch.

Plate XI, Figs. 1 to 4.

The disease which is popularly known as gid. sturdy, staggers, or turnsick in sheep, is caused by the presence of a parasite living in the brain. This parasite is closely related to *Tænia marginata*. It lives in about the same way, but differs from it in detail. It is known as *T. cœnurus* in its adult state, and as *Cœnurus cerebralis* in its cystic state—the stage in which it infests sheep.

Method of infection.—Sheep become infected with this dangerous parasite while pasturing or drinking where dogs have scattered the eggs in their wanderings through the country. In the West the eggs may also be scattered by the wolves, coyotes, and foxes, which may harbor the adult parasite.

Life history.—The eggs of *Tænia cœnurus*, after being taken with the food or drink, are hatched within the stomach of the sheep and make their way through its walls. They then migrate either actively, by forcing their way through the connective tissues, or passively, as is generally believed to be the case, by getting into the circulatory system, and are carried from thence into various portions of the body. Those arriving in the spinal canal and cranial cavity seem to thrive and grow, while the others, which may have reached the heart, lungs, and diaphragm, grow for a time and then perish.

Description of cystic stage.—Having arrived in the brain cavity, the young embryo migrates upon the surface of the brain much as the embryo of *Tænia marginata* does through the liver substance. The galleries it makes are sinuous. They begin at a point and slowly increase with the growth of the parasite and run in any direction. In from two to three months after the first invasion of the brain the cysts have grown as large as a hazel-nut, or about a half inch in diameter. When examined closely they will be found incased in a thick outer skin, a sac made by the membranes of the brain. Out of these sacs the parasites may be loosened, and these resemble the cysts of *T. marginata*. It differs, however, in a very important particular—instead of having a single knob attached, tens or hundreds of these knobs may be seen as little dots hanging from the inner surface into the fluid of the cavity, (see Figs. 3 and 4). Each of these dots can evert itself, or push its head out, and will then be seen by the aid of a glass to be a perfectly de-

83

veloped head, having four suckers and a crown of about twenty-eight hooks. These heads, when the cysts are fed to dogs, may develop into as many individuals. Most of them will generally die, and only a few of the stronger will develop. Instead of the single worm which the embryo of *T. marginata* produces, this peculiar species develops many from each of its embryos. In this there is a compensation; for while many of the *T. marginata* embryos come to maturity, but one or two of the *cœnuri* survive, and thus the opportunities for the further perpetuation of the species are diminished.

Duration of development.—Experiments have shown that the embryo may be found in the brain from two to three weeks after feeding, and is then about the size of a mustard seed. Between three and six weeks after feeding the worst symptoms of the disease occur. The *cœnurus* becomes developed in from two to three months. After this time it continues to grow in size and in the number of heads for six or eight months, when it usually causes the' destruction of the affected sheep. When the developed *cœnurus* is fed to a dog it usually produces adult tape worms within a month.

In the migrations of these parasites many are lost and destroyed. Of the hundreds of eggs which leave the intestines of dogs few reach the stomach of the sheep, and of these still fewer enter the cranium. Of the few which become adult—one or two in each sheep affected—but a small percentage nowadays arrive in the dog again. Thus of the thousands of embryos that left the dog but a single *cœnurus* may return. But this *cœnurus*, developing again into several tape-worms, is the source of many new embryos for the re-infection of the sheep.

Disease.—Lambs and yearling sheep seem to be most liable to infection; those over two years old seem to possess a degree of immunity. Sheep herded by dogs; those breeds which eat the grass close to the ground; sheep which drink out of ponds or brooks in which the dogs bathe; flocks belonging to careless flockmasters, who leave the skulls and viscera of slaughtered and dead sheep strewn through the pastures, are more liable than others. In short, any of the conditions which help in the distribution of the parasites render sheep more subject to the disease.

Symptoms and progress.—The symptoms of gid in sheep are dependent upon the stage of invasion and development at which the parasite has arrived. The invasion embraces the period from the time that the embryos have been swallowed to the time that they become partially developed in the brain. The stage of development embraces the remaining time they pass in the brain. The stage of invasion generally passes unnoticed. Between the second and third week those animals worst infected—for but few of those infected show signs of disease in this stage—show symptoms of inflammation of the brain and surrounding tissues. It is at this period that the little parasites are active in progressing through the tissues. Dullness, feebleness, heat in the head, intense redness of the mucous membrane of the eyes, and increased pulse-beat are characteristics of this stage. The head is generally held in a peculiar position, either stretched at length, turned backward, inclined to either side, or drooped. The intensity of these symptoms depends on the number of invading parasites. Later, spasmodic convulsions or paralysis may occur. Death may take place in about six or eight days after the first symptoms are noticed. The diagnosis at this stage is difficult, but depends on finding the parasites and their channels on the surface of the brain. The former are at this time the size of a mustard or flaxseed. A correct diagnosis at this stage will influence the future treatment of the flocks. If the sheep live through this stage no further symptoms will be noticed for from four to six months, when a new set of symptoms will appear. It is said that of all infected sheep less than 2 per cent. of those which show the disease in the early stage recover.

The symptoms of the second stage arise from two causes—from the irritation produced by the little heads thrust out of the mother bladder and from the disturbances created by the pressure caused by the increase in size of the cænurus.

The actions from which the disease has derived its common name in all countries are especially characteristic. The head turns; the animal walks in a circle; it staggers, trembles, has convulsions, acts stupid; it becomes unable to eat or drink, and finally dies of exhaustion or paralysis. The various gaits and peculiar positions assumed by the infected animals arise from the particular location of the parasite. The length of time between the attack and death also depends on this position, some parts of the brain being more vital than others. Death occurs in a month or a month and a half after the secondary symptoms appear.

A *diagnosis* of the disease in this stage can be determined by an examination of an infected animal. Sometimes at this stage the parasite softens the wall of the skull by its continued pressure and the spot can be felt with the fingers. *Cænuri* in the spinal canal are more difficult to diagnose. They cause the sheep to stagger and walk peculiarly with the posterior extremities. Sometimes the disease is manifested by an intense itching along the backbone without any apparent reason. Paralysis of the hind limbs and of the rectum and bladder often result.

The gid should not be confounded with the disease caused by grubs in the nasal cavities, which sometimes produce similar symptoms. The false gid produced by the larvæ of the *Œstrus ovis* will cause more symptoms of a catarrh or cold in the head and less of the turning, which is so very characteristic of true gid.

Treatment.—There is no treatment that can be profitably followed for sheep thus affected. A veterinarian could in the more advanced cases locate the position of the cyst either by inference from the character of the symptoms or by feeling the softened spot in the cranium. He might, by carefully cutting away or trephining the skull over the point, remove the parasite and enable the sheep to recover. The intelligent farmer might learn to do this himself, but after it is all done the cost of doing it will about equal the value of the sheep saved. The true treatment, and that which has saved France and Germany more than any medicinal or surgical treatment devised, consists in prevention.

Prevention lies in the treatment of the sheep and of the dogs. As the developed *cœnurus* comes only from the cranium or spinal canal, it is very easy to prevent dogs from being infected by taking care that they can not get these portions of the carcass either when the sheep are slaughtered or after they have died in the pasture or sheep-cote. The heads should either be rendered, burned, or deeply buried, and not thrown into the first convenient corner.

When the skulls and viscera must be fed to dogs they should be subjected to a prolonged boiling. The soup so made would be harmless. When the lambs are known to have passed through the first stage of the disease and are fat enough for sale, at the very beginning of the secondary symptoms they should be slaughtered and marketed, care being taken with the first killed to verify the diagnosis. This will save more than any other proceeding. The treatment and handling of the dogs are the same as for *Tænia marginata*, except that *T. cœnurus*, according to Leuckart, develops in three or four weeks, and has to be medicinally attended to oftener, or until the dogs are quite free from it.

The adult.—*Tænia cœnurus*, the adult tape-worm, which grows from the cyst and causes the gid when in the sheep's head, resembles *T. marginata* and also *T. serrata*, a tape-worm which the dog acquires by eating the viscera of rabbits in which the young form is encysted. *T. serrata* is about as large and long as *T. marginata*. *T. cœnurus* is much smaller than either, measuring when mature between 1 and 2 feet, while the former measures a yard. It is also a slenderer species. The most decided differences lie in the hooks of the head. *T. serrata* has the largest head, the largest hooks, and the largest suckers, the latter being a third larger than those of *T. marginata*. They are from thirty-eight to forty-eight in number. The hooks of *T. cœnurus* are between twenty-four and thirty-two in number, and the slenderest of the three species. The terminal segments also vary, those of *T. cœnurus* being the smaller. The total number of joints also differ, *T. marginata* having five hundred and fifty or six hundred; *T. serrata*, three hundred and twenty-five or three hundred and fifty, and *T. cœnurus* about two hundred. Exact measurements of all these parts as given in text-books enable one to definitely determine the species, but the flock-master needs most to know that his own and his neighbor's dog harbor tape-worms, which are prejudicial to his flocks, and to proceed against them.

The presence of *Hydatids* (*Tænia echinococcus*, v. Siebold), (Plate XI, Figs. 7, 8, and 9), is, if it occurs at all in this country, very rare. It has a life history similar to *T. marginata*, passing from sheep, cattle, and pigs to dogs, and from dogs back again. In method of growth it differs, forming from a single embryo large masses of cysts, which contain many individuals capable of becoming adult. Its favorite situations in sheep are the liver and lungs. It may occur elsewhere. In these places it forms large bladdery masses, whose nature can only be definitely determined by the aid of the microscope. As this parasite is also dangerous to man the bladders should be handled with care, and destroyed. The treatment of dogs is the same as for the other *tænia.*

Tænia tenella, Cobbold (Plate XI, Figs. 5 and 6), which causes *mutton measles,* is a tape-worm of man, and is supposed to be acquired by him while eating mutton through which the *cysticerei* have scattered. The disease has not yet been noticed in this country, and having been observed but seldom in Europe, is yet problematical. Mutton affected would present little white spots as large as flaxseed scattered through it. The loin muscles are most apt to be infected. Neither this disease nor *hydatids* can at present be accounted a disease of sheep in this country.

TÆNIA CŒNURUS, Küch.

PLATE XI

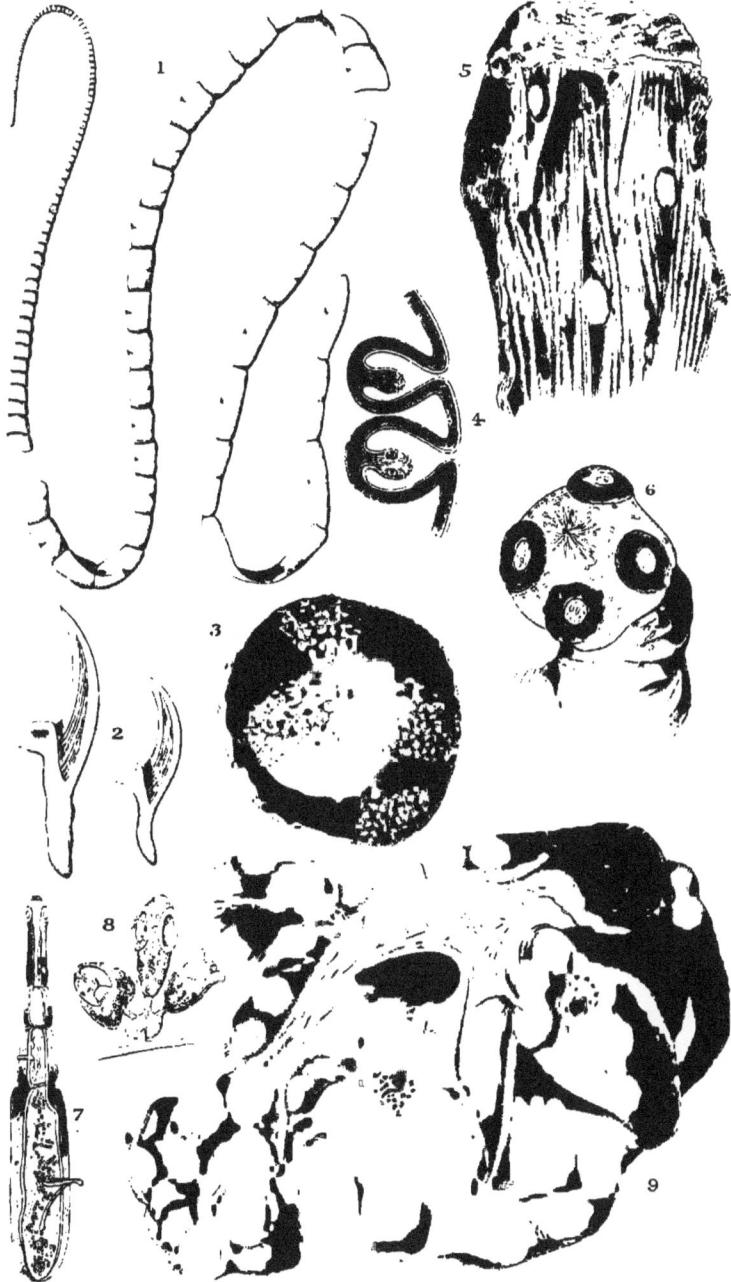

Haines, del.

TAENIA COENURUS.   CYSTICERCUS OVIS.

TAENIA ECHINOCOCCUS.

ADULT TAPE-WORMS.

Plates XII to XV.

The flocks of this country are infested by two species of adult tape-worms, remarkably different in form, size, habits, and distribution. They are *Tænia fimbriata*, Diesing, and *T. expansa*, Rud. Each are named from predominant characteristics.

Neumann states that European sheep harbor more species of adult tape-worms than other animals, and enumerates ten species which have been described, viz: *Tænia expansa*, Rud; *T. alba*, Kerr; *T. Benedeni*, Moniez; *T. aculeata*, Kerr; *T. ovilla*, Rivolta; *T. Giardi*, Moniez; *T. Vogti*, Moniez; *T. centripunctata*, Riv.; *T. globipunctata*, Riv.; *T.ovipunctata*, Riv.

The majority of these species are apparently new to science, and consequently unconfirmed. Some seem to be well founded, while others may eventually prove to be re-descriptions of other better established species. With the exception of *T. expansa*, the writer has as yet found no traces of any of them, and it is probable that they do not occur here.

A description of the tape-worms found here, and the diseases they cause, follows.

THE FIMBRIATE TAPE-WORM.

TÆNIA FIMBRIATA, Diesing.

Plates XII and XIII.

Tape-worms in Western sheep were noticed by ranchmen in the early days of ranching, but did not attract the notice of veterinarians until 1883–'84, when Dr. Faville, of the Colorado State Agricultural College, first directed attention to them. (Report Veterinary Department of Colorado State Agricultural College, January, 1885.) An earlier epizöotic, due to tape-worms, had been reported to Mr. Stewart, who, in the National Live Stock Journal, for September, 1875, records their presence in Missouri sheep, and from specimens at hand determined them as *Tænia plicata*. As this *tænia* does not occur in sheep, but in horses, it is quite likely that Mr. Stewart saw *T. fimbriata*.

In a recent letter to the Department of Agriculture, the late Hon. J. M. Givens reiterates an opinion formerly expressed by him in local societies of wool-growers, and published by him in Denver (Colo.) papers of 1883–'84, that these tape-worms were a cause of the larger part of the losses among sheep, and urged the necessity of a closer study of the subject, in order that more might be learned of the parasite, the amount of loss it caused, and the means of preventing it. These losses had previously been attributed to a weed called "loco," which the sheep ate.

In obedience to instructions received from the Commissioner of Agriculture, the writer proceeded to Colorado in August, 1886, and began a study of the various intestinal parasites of sheep. The studies of that year were pursued, by invitation, on the ranch of Mrs. Mary L. Givens, who, at great inconvenience to herself, did all in her power to aid me. In the spring of 1887 I again visited Colorado Springs and prosecuted other experiments, which it was hardly possible to conduct in the East.

Early in my investigations the fact was recognized that one *tænia*, identified as *Tænia fimbriata*, Diesing, was the most abundant; that it was scarcely ever absent in sheep examined, and was the probable cause of most of the tape-worm disease of Western sheep. As this *tænia* was so constantly present attention was directed to it, and an endeavor made to obtain it in all of its stages, and to learn how and where it passed its entire life. Another fact was soon learned, that the other species of tape-worm (*Tænia expansa*), usually abundant in lambs, was found so infrequently that it was difficult to find sufficient material for study. The methods of study were, first, observation, and, second, experimentation. The experiments have, as yet, been productive of nothing but negative results. In determining so much of the life history as has been learned *post-mortem* observations have been the most useful. Not only have animals been slaughtered on the ranch for this purpose, but advantages offered by inspections of sheep slaughtered at the shambles for consumption have been utilized. From these inspections the abundance of the parasite, the first appearance in lambs, the different stages in growth, etc., could be learned, but it soon came to be necessary to study the surroundings of the sheep—the corral, the watering places, and the range—to learn under what condition the parasite must exist while external to the sheep, i. e., while passing from sheep to sheep.

The effects of the parasite on its host (the sheep) were arrived at from studies of the flocks and from observation of individuals at *post-mortems*. The changes found were noted and careful attention paid to the point whether certain changes were due to parasites or another disease called "loco." As all of the sheep examined were called "locos," it is evident that there was here a source of error. No certain pathognomonic lesion of "loco" was learned; as all of the "locos" were infected with *tænia*, the separation of characters due to each disease was difficult. Indeed, it has seemed to me that all of the symptoms due to the parasitic disease may be ascribed to "loco." The characters of the tape-worm disease are, however, well marked in lambs which have never learned to eat this weed exclusively, and thus could be studied without regard to "loco," which, if a disease at all, belongs to yearlings and older sheep.

Experiments were undertaken with a view of infecting lambs with the parasites, while the lambs were kept from other possible sources of infection; but these were fruitless. Other experiments were entered upon with a view of removing these parasites by medicinal remedies. None of these were effective in removing the parasites from the gall-ducts, and were abandoned until more could be learned of the life history of the parasite, when they could again be taken up with perhaps a better prospect of success.

· The total results regarding this *tænia*, so far obtained, are as follows, viz:

The parasite persists in an adult stage in the older sheep throughout the year.

The smallest forms appear in lambs soon after the second month of their age, and may be found in sheep of any age throughout the year, excepting, possibly, the winter months.

It requires at least six months, possibly ten, to attain an adult size.

The ova or embryos are continually passed from the sheep to the ground throughout the year. The life of the embryo from the time it leaves one sheep until it is found in another is yet undiscovered.

When present in considerable numbers in sheep it determines a disease which is not only detrimental to the value of the animal, but at times causes the death of large numbers.

No medicinal remedies can be recommended which will assuredly remove the parasite from the host.

Many measures may be taken which may prove to be effective in two ways, first, in preventing sheep from becoming infected; second, in enabling the sheep to better withstand the ravages of the parasite, and thus carry it over the critical stage of its existence.

The lambs and yearlings are the greatest sufferers, and it is to these that the most attention must be paid.

TÆNIA FIMBRIATA, Diesing.

SYNONYMY.—*Thysanosoma actinioides*, Diesing, 1834. Med. Jahrb. d. öesterr. Staat. Neue Folge, VII, 105-111. Taf. III (Fig. inverse delin.), Ej. Syst. Helm. I, 501 in nota.

Tænia fimbriata, Diesing. Syst. Helm. I, 501.

Tænia fimbriata, Diesing, 1856. Zwanzig Arten v. Cephalocothyleen, p. 11, 1856. Wien.

Tænia fimbriata, Rept. Dept. Agric., 4th and 5th Annual Repts. of Bureau Animal Industry, 1887, 1888, p. 167, Pls. I and II.

Tænia fimbriata, Diesing, was first discovered by Natterer, in Brazil, in 1824, and published by Dr. K. M. Diesing in 1834, as a new genus, *Thysanosoma actinioides*. Later, in 1856, Dr. Diesing republished this parasite as *Tænia fimbriata*. (See Plate XIII.)

The first specimens found were detached segments, and from these the first description was made. Natterer later found more complete specimens and upon these the species was founded. The specimens were found in the intestines of the following species of deer, viz: *Cervus paludosus, C. rufus, C. simplicornis, C. Nambi, C. dichotomus.*

A translation of the original Latin description is as follows: *Arhynchotæniæ,* Rostellum none; mouth unarmed.

Tænia fimbriata, Diesing. Tab. V, figs. 9–15. (Plate XIII is a copy.)

Head obtuse tetragonal, large with hemispherical angular bothria; neck none; body anteriorly lanceolate, with very short cuneate segments; posterior margin of the upper entire, of the following crenate, of the last fimbriate on each side; the linear fimbriæ rounded at the apices. Genital apertures —; length 6''' — 6''; width 1 — 3'''. Mature segments, separate, 1''' long, 2''' wide, with lanceolate fimbriæ.

This description was necessarily imperfect, from the lack of abundant material, but is nevertheless sufficient to enable us to identify the present species with it. Fortunately, too, excellent figures enable us to clearly understand the fimbriate character of the species. At present this is the only species known to possess this character. Though characters founded on form are of doubtful value, until more decided differences are determined between the fimbriate *tænia* of deer and those of sheep, this decidedly strong character must serve to keep the two together. Besides, there are no strong reasons why the two should be separate, for each is of about the same size, a fact which we would scarcely expect in the same parasite living in hosts of different genera; and each is also found in hosts of similar body temperatures, habits of life, and of feeding. That they are found in far-separated localities need be no serious objection, for the land connection of North and South America would permit of the infection of deer of both continents with the same species of parasites.

Tænia fimbriata is lanceolate when contracted, linear when relaxed. It is quite thick, the fimbriæ on the contracted specimens presenting the appearance of plush. The segments can only be distinguished on the more relaxed specimens. Adult specimens are from 15 to 30^{cm} in length and about 8^{mm} in width. Immature specimens range from less than 5^{mm} upwards. The greatest width is about 2^{cm} from the free end, from which point the segments become narrower. There are at the free end of adult specimens from three to four or more segments which are of nearly equal width, which have lost their contractility and are in the process of being shed. The shedding of segments begins in youngest specimens and continues throughout the life of the parasite.

The head or organ of attachment is depressed and tetragonal and quite large, sitting on the neck like a pin-head; it is from 1 to 1.5^{mm} wide, hookless, and has four very large suckers. The substance of these cups forms the greater part of the head.

The neck or the part where segmentation begins is very short in contracted specimens, but can be seen in the relaxed condition. The seg-

ments are very short and flat near the head, but concave or cup-like
toward the free end, each overlapping the succeeding, and all appearing
linear on the surface of the *tænia*. The terminal relaxed segments are
cuneate. The borders of segments nearest the head are slightly wavy
or crenate. They soon become fimbriate even in youngest *tænia;* so
that the smallest specimen found demonstrated the fringed character
of the species. The fimbriæ may either be contracted when they are
stout and short, or relaxed when they are flaccid and proportionately
longer. They are obtusely pointed. The segmentation in contracted
specimens is made out with difficulty. It is indicated by transverse
striæ. In relaxed specimens it is plainer.

The sexual organs are symmetrically placed, two sets in each segment, each opening in a lateral pore. Each set of organs is composed
of a male and female portion, or is hermaphroditic. They begin to develop at some distance from the head and attain maturity towards the
middle of the *tænia*.

Besides being remarkable in the fimbriate character of its segments
this species is also peculiar in the form of its reproductive apparatus.
The male portion develops first and occupies the whole width of the
young segments. It consists of sacs connected by tubules with a large
tube which finally becomes the much convoluted efferent tube.

The ovaries develop later, and are situated at each side of the segment. They are not shown in the plate.

The uteri develop last. Each is made of a series of bags arranged
side by side in a fringe which extends along the top of the segment
from side to side. These bags open into a larger tube from which they
receive the developing embryo. The tube connects with the ovaries.
The embryos develop in the uteri, and probably remain there until the
segments go to pieces on the ground and thus permit them to be scattered. They are six-hooked and not essentially different from those of
other *tænia*.

Occurrence.—This *tænia* is found in the duodenum and gall ducts of
sheep. The former is sometimes found containing from thirty to one
hundred specimens. More often, however, there are from two to thirty.
The gall ducts are frequently completely distended by the *tæniæ*, which
pack them so tightly that the parasites can not be withdrawn by pull-
ing without breaking. Occasionally one, or at the most two, may find
their way into the pancreatic ducts, which they also distend. They get
into these ducts when young and distend them as they grow larger.

A few disjointed segments may be found below the duodenum, but
no entire individuals. Nearly every sheep of a flock will be infected.

Distribution.—*Tænia fimbriata*, Diesing, the fringed tape-worm, is at
present the most common parasite of the sheep of our Western plains,
and causes by far the greatest loss of any intestinal parasite in this
country.

As may be seen by inspection of the tables showing parasites found in different *post-mortem* observations, it has been identified in sheep from Utah, Colorado, and Nebraska. Mr. Codweis, of Granger, Colo., a former sheep-owner in New Mexico, says that he has seen them there. Mr. Samuel Collins, of Colorado Springs, Colo., who has slaughtered sheep from California, Oregon, Utah, Nebraska, and Colorado, says that all sheep from these States have them. Dr. Faville personally told me that he has seen them from Oregon sheep and from sheep in New Mexico. Mr. Stewart's identification of *Tœnia plicata* from Missouri sheep (National Live Stock Journal, September, 1875) leads me to suspect its presence in that State. When to these evidences of wide-spread distribution we add those offered by the intermingling of Western sheep by parentage and traffic, and by the opportunities for infection presented by the nearly unrestricted communication of the ranges, we may believe that this distribution is necessarily wide-spread. Its distribution at present is from Oregon and Wyoming southward and from Nebraska and Missouri westward, or, more accurately, from the ninety-fifth degree of west longitude westward and from the forty-fifth degree north latitude southward. It coincides with the distribution of the sheep in those parts, and more especially with that of the descendants of the Mexican or Spanish sheep with which nearly all of the larger ranches were originally stocked. The precise limit of its Eastern distribution is unknown, but is probably limited to those portions of Nebraska, Kansas, and Missouri to which Colorado feeders have been sent prior to selling them to the Chicago markets.

There are at present no facts at hand to show that the parasite exists east of the Mississippi River. In two instances a number were found in sheep slaughtered in Washington, D. C., but these animals were said to have come from Chicago, Ill.

Life history.—All of the life history of this *tœnia* that is at present known has been learned from *post-mortem* dissections and microscopic investigation.

The adults were found in yearlings and older sheep throughout the year. No adults have yet been found in lambs less than ten months old. The smallest stages of the immature *tœnia* may be found in all young sheep over ten months old. They are usually most abundant in lambs, yearlings, and two-year olds. Although a sheep may be infected with a number of *tœniæ* of about the same size, indicating an infection covering but little space of time, it is more usual to find the parasites of various sizes, indicating a continuous infection. The retention of food and liquids for some little time in the rumen and reticulum may account for this in part. These varying sizes continue from May until December. Sheep examined in May presented various sizes and indicated infection in former months. No other data showing infection during winter months were obtained. The smallest *tœniæ* are found in the duodenum; those found in the gall ducts are larger. *Tœnia* less than 2^{mm} long have been found in the duodenum after the gall duct has become completely packed by the parasites. The adult worms have embryos in their most distant segments, which are ready to be set free. These embryos escape from the host with the feces. Until they reappear in the duodenum of another sheep, a quarter of an inch in length, their history is unknown.

A *tænia* infecting a lamb two months old, the youngest stage noticed, is about a half inch long ; as the season advances it is joined by others, and these increase in size. Four or five months afterward it is found to be 4 or 5 inches in length, showing a monthly rate of growth of about 1 inch. From this time it gradually increases in size until the following spring, when it becomes adult and capable of furnishing embryos for infection of other animals. These embryos escape from the sheep, and while many are destroyed a few arrive at their destination in a second animal.

Disease.—The influence that the presence of *Tænia fimbriata* has on the life and health of its host is not inconsiderable. The ultimate loss is seen when lambs which should be fat and strong are not, and die during the colder weather while the fatter ones survive. This loss, where the hosts do not die, can not perhaps be accurately estimated, but is nevertheless present, for thin, hide-bound, dwarfed sheep are not valuable for mutton, nor do they produce as much wool as they otherwise would.

So slowly are the parasites hatched, so slowly do they grow, and so gradually do the symptoms develop, that the *tæniæ* are present in considerable numbers and size before systemic disturbances in the lambs present themselves. An experienced ranchman will probably notice towards September that some of the lambs are not growing as they should. Later in the fall the symptoms increase. In November the lambs, which are by this time thoroughly infected with a number of strong, tenacious parasites, show the disease quite plainly.

The disturbances finally shown are induced at first by the local irritation produced by the worms attaching themselves to the villi of the intestinal walls and causing a greater secretion through their strong vermicular action. A continuance and increase of this irritation caused by the growth of the parasite and an accession of other parasites, finally excites chronic catarrhal inflammation of the duodenum and biliary duct. To these disturbances we must add those arising in the liver from a plugging of the duct by the parasites, which grow so large that they distend it to a comparatively large size.

Dr. George C. Faville, in a report of the veterinary department of the State Agricultural College of Colorado for 1884, describes the *post-mortem* appearances of these animals as follows :

Organs of thorax were normal. In the abdominal cavity I found the stomach filled with a mass of semi-digested loco leaves. The liver was normal in appearance ; gall bladder filled with greenish-colored bile. In the duct running from the gall bladder to the small intestines, I found a mass of tapeworms (*Tænia expansa*). The small intestine I found filled with a mass of these worms, varying in length from 6 inches to

5 or 6 feet. The kidneys were normal in size and color, but upon section, I found the pelvis filled with a gelatinous material. The muscular system was exceedingly flabby and pale in color. The body seemed to be absolutely destitute of fat. The urine was normal. The brain showed a slight serous effusion about the base, and to a slightly greater extent in the region of the medulla oblongata. There also was a slight effusion into the abdominal cavity. The only other change that could be found in the brain of these sheep was a slight congestion of the arachnoid membrane.

The above description is taken from so-called "locoed" animals, but applies equally well to tæniæ-infected sheep. Of the many "locoed" animals examined, but one or two have been free from *tæniæ*, and in these the gall ducts were thickened and enlarged as though they had at some earlier date been infected. It is extremely difficult to separate the symptoms of the two diseases, and it seems to me that many cases of "locoed" animals are victims of the tapeworm. That the *tænia* may tend to produce depraved appetites and the morbid craze for a particular food, is also a reason for suspecting that the loco disease may depend in part on the tapeworm disease.

In Dr. Faville's description there is one point which deserves attention, and that is the finding of a slight congestion of the arachnoid membrane. In specimens examined by the writer there seemed to be no undue congestion, and the arachnoid membrane, which is a vascular one, naturally looks red or dark colored. The brain symptoms of these animals are such as arise from anæmia rather than hyperæmia of the brain.

In affected yearlings which are not suspected of eating loco more than other animals (all eat of the loco plants), the following *ante* and *post-mortem* symptoms can be observed: Lambs that are badly affected are large headed, with undersized bodies and hide bound skins. Their gait is slightly like that of a rheumatic. They seem to have difficulty in cropping the shorter grass; they also appear to be more foolish than the other sheep, standing oftener to stamp at the sheep-dogs or herder than the healthier ones. Others do not seem to see as well, or are so affected that they appreciate danger less. In driving they are to be found at the rear of the flock. Internally the organs present no marked symptoms of disease. The abdomen contains more dark-colored serous fluid than normal; the omentum is often nearly devoid of fat. The catarrhal inflammation and thickening of the mucous membrane of the duodenum and gall-ducts have already been noticed. The liver, in cases of long standing, is somewhat smaller than normal; the kidneys are sometimes flabbier and paler than normal; the lymphatics look somewhat darker; the muscles are thinner and weaker. There is in all cases a diminution of fat, and in most cases the leanness of muscle is marked. In those places where the fat usually occurs in masses little or none is found. Associated with this condition is the presence of serous infiltration of the connective tissue in the abdomen, thorax, spinal and cranial cavities. This infiltration is the most marked in the worst cases. The groin, the pelvis of the kidney, the spaces between serous coats of the abdomen, and other spaces where serous membrane partially or entirely surrounds an organ are noticeably infiltrated.

These conditions hold in lambs and older sheep. Between the worst affected and entirely healthy individuals there are many grades. The symptoms and pathologic lesions are those of mal-nutrition, and aside from the lesions of the duodenum and liver are not materially different from the systemic lesions caused by other parasites, or from innutritious food, or from any cause that would prevent the animals obtaining and assimilating nourishing food. A variety of other causes would produce the same lesions.

The parasites may produce their evil results as follows: Their vermicular actions cause increased secretion of the intestine where they are lodged, both by direct irritation and sympathetically, i. e., the adjacent intestine secretes more than it ordinarily would by acting in sympathy with the infected portion. This hypersecretion soon becomes abnormal, and the secreting membranes become so changed that they can no onger act physiologically. Its best purpose is in furnishing the parasite with more nutritious fluid. The plugging of the gall ducts not only stops the gall from flowing at proper times, but dams back that which is secreted during digestion, and allows it to slowly ooze out after it is needed. When the ducts are unobstructed the bladder and ducts are lemptied at their proper times, and any interference with this flow deranges healthy digestion. The damming back of the gall reacts on the secretion in the smaller ducts, and this in its turn reacts on the physiological functions of the liver cells.

The disturbance of digestion due to this impairment of the functions of the liver and duodenum has not a merely local effect. In the upper parts of the small intestine important digestive changes take place, and the disturbance of any of these prevents the proper preparation of food for its assimilation through the intestinal wall, resulting in a loss to the animal of nutrient material. The duodenum is held to be a very irritable organ, diseases in it causing reflex disturbances of various kinds. These reflex actions also lead to many systemic disturbances. Now these disturbances are each slight, but when combined and continued through weeks and months they cause the results just described. To one seeing a half dozen or more worms taken from the intestine of a sheep the worms do not seem to be a sufficient cause of disease. The disturbance caused by one worm in man gives rise to even greater systemic derangements. The non-assimilation of food and reflex irritation produced by the tape-worms seem to be the chief causes of the impoverished condition of the infected animals. From these causes proceed the imperfect nutrition of the various organs and the dropsical effusions resulting therefrom.

From this state of mal-nutrition all of the systemic disturbances can result. The staggering gait may arise from the weakened muscular system; the absence of fat from non-deposition of more and the consumption of that heretofore deposited; the serous effusions from the weakened condition of the system; and the foolish actions from the long-continued lack of nourishment of the brain.

Sheep do not die from the tape-worm disease alone. The greatest losses are, the ranchmen say, among the lambs and yearlings. The majority may die during cold storms, either from freezing or from suffocation while piling upon each other for warmth. They may starve to death either from inability or lack of desire to eat. They may die from other diseases. The tape-worm disease appears to render them more liable to other affections and less able to withstand the inclement season. It is therefore indirectly chargeable with the loss. Even if the infected sheep do not die, the parasite is still a cause of pecuniary loss. The impoverished condition traceable to it is a small average loss for each animal, but for flocks of over five thousand sheep the aggregate is thousands of dollars for each ranchman.

In the article of Dr. Faville, cited above, he quotes a letter * from the late Hon. J. M. Givens, whose flocks numbered from six to eight thousand head. In this letter Mr. Givens states his loss from dead sheep alone for the preceding year at from $3,000 to $4,000. Fortunately the loss of from four hundred to eleven hundred or more sheep does not occur to flockmasters annually, but such losses are not infrequent, and may be heard of either on this or that ranch during different years. Every ranchman knows of and appreciates the steady though small loss arising from the depreciated value of his animals, due to their ill condition from various causes, and which he strives by every means to reduce, for therein lie the profits and success of his business. From the study and observation which the writer has been able to devote to the tape-worm disease it appears alone responsible for more losses than any other sheep disease on the prairies excepting scab. The direct death-rate traceable to it is large when compared to the entire death-rate, and the indirect loss traceable to it is, though more insidious in its character, still larger, for it is ever present and ever active.

Medicinal treatment.—Some experiments looking toward the removal of *tænia* by medicines were made in 1886. Various *tæniæfuges* were tried with little success. The powdered preparations of pumpkin-seed; pomegranate-root bark, koosoo, kamala, male fern, and wormseed proved of no avail.

In order that they might be administered cheaply the proper amount of each for ten animals was mixed with meal, bran, and salt, and fed in

---

*The letter referred to gives "loco" as a cause of the losses. Before his death the Hon. J. M. Givens had concluded, and communicated to his friends of the El Paso Wool Grower's Association, that the loss of this winter was not due to "loco," for the greatest loss had occurred in young sheep and lambs. The latter had not learned to eat "loco" exclusively, were poor, and presented symptoms which he learned later belonged to sheep infested with tape-worms.

a trough. When sufficient meal and salt was mixed with the medicines to entice the sheep to eat it, the bulk that contained the requisite dose of medicine was too large for a sheep to eat at once. As this bulk was retained for some hours in the rumen, the efficacy of the dose was lost, for the virtue of nearly all of these remedies depends on the dose passing through the intestines in mass. Human patients are usually prepared for the medical treatment by abstaining from food for at least twelve hours previously; they are then given a cathartic which is followed by the anthelmintic. This plan of treatment utterly fails in ruminants, for neither stage can be successfully carried out in administering these remedies by the mouth. The presence of the large rumen, which holds a large quantity of reserve food, and into which new material may be taken, accounts in part for this. Some of the food, if sufficiently fine, in fasting animals passes directly to the manifolds and fourth or true stomach, but a certain proportion would fall into the rumen, and thus the efficacy of the total amount acting within a given time would be lost.

These experiments failed, therefore, through the anatomical structure of the animal and the method of administration. The presence of *tæniæ* in the biliary ducts is another reason why *tæniæfuges* can not be entirely successful in treatment of sheep with *T. fimbriata*. Any medicine which would affect the *tæniæ* in these ducts would also affect the sheep seriously. It is doubtful whether they can be killed or driven from the ducts. The continued or repeated administration of remedies that are necessary for expelling these *tæniæ* is also an objection to their use. The parasites are continually appearing throughout the year, and even if those already developed could be driven off, the constant re-infection would necessitate other operations for their removal. The cost of the necessary medical treatment seemed to me to more than exceed the good results that possibly might be realized. Further experiments were therefore delayed until the complete life history of the parasite should be determined. In this history we may hope to find some stage at which we may more profitably administer remedies. Many prescriptions for eradicating tape worms in ruminants are given in various journals and agricultural papers. Some of them when tried may have proved very efficacious. Unfortunately reports concerning the effects of their administration are not recorded. My own experience leads me to have little faith in them. There is a feature about them which, no doubt, has been recognized by the ranchman who has undertaken to carry them out to the letter, viz: It is the entire inadequacy of the recipes in prescribing methods of administration, and medicines of reasonable price as well as of certain efficiency. This oversight is of such importance that otherwise good recipes have to be abandoned. The Western methods of treating sheep medicinally must differ from the Eastern methods, as the methods of sheep-dipping, sheep-shearing, or sheep-husbandry in these sections differ; otherwise the expense of treatment will be so considerable that in view of any uncertainty of cure few ranchmen will undertake it.

Preventive treatment.—The most effective service rendered to man and beast by the physician has been through the prevention of diseases and the preservation of health by hygienic measures. Appreciating this, and that effective prophylactic treatment of the sheep against infection by *Tænia fimbriata* could not be realized except by the most thorough knowledge of the complete life history of the parasite, my attention was turned to the investigation of its younger stages and those of other unarmed *tæniæ* which were available. Such are the difficulties of this investigation that the gap in the life history, which may exist between the time when the embryo passes from the sheep until it is found, less than a quarter of an inch long, in another sheep, has not been completely investigated.

From the present knowledge of the development and life of this parasite there have arisen more difficulties in forming rules of prevention than was at first anticipated. The presence of the adult and young parasites throughout the year, and the methods of Western sheep ranching are factors which are all-powerful in keeping up the tape-worm disease. The case is not a hopeless one, however, for there are certain phases of feeding and watering the sheep which can probably be advantageously changed, both for the prevention of this and other diseases.

The feeding occurs on the prairie and in the corral. I would recommend that the ewes with their lambs should be pastured on a portion of the prairie that had not been run over by sheep for some months previous. They could be driven to the new pastures about the time that the lambs begin to nibble at the grass and drink water. After the lambs are weaned they should be changed to fresh uncontaminated pastures until winter, and other older sheep put on the range vacated. If there be sufficient range the lambs could be kept on as nearly uninfected ranges as possible until they become two-year olds. In feeding lambs on grain and hay measures should be taken to keep the food from the ground. The grain should be fed from troughs placed either on a board floor that could be cleansed or on ground kept scrupulously cleaned of all droppings. The hay should be fed from racks. The corrals for the lambs should either be fresh ones or the old ones should be periodically and thoroughly scraped out and cleaned. They should not be put with a greater number of old sheep than is absolutely necessary.

The watering occurs at various places. The usual method is the watering at rivulets or ponds. This should be done, but such places should be fenced in and troughs provided into which the fresh water could run. These troughs should be raised a little above the surface of the ground so that they could receive no surface drainage. By the aid of pumps and wind-mills this could be easily accomplished. Most watering places are so situated that by conducting the water through pipes or boxes but little expense would be necessary to guide it into troughs. These troughs should be kept clean. The lambs should not be allowed to drink elsewhere, nor to eat grass in moist places unless it is absolutely certain that these places are uninfected. The location of corrals so that they either surround water or that the drainage is from them to the water seems to be a most harmful practice. It not only makes the water fouler but renders it more likely to hold parasites. Herders should be instructed neither to feed nor water at the prairie pools. If there are places where it is advisable to water they should be prepared like the watering places at the corrals. The nearer the ranchman can arrive at giving the lambs pure fresh water the less infected with parasites will they become. The salt for the lambs and young sheep should be fed from boxes placed near water places and kept constantly full. They will take no more than they want and will be all the healthier if they have all they need. If they are deprived for a time they may at first salting eat more than is good for them. A little eaten daily is physiologically better than the larger quantity eaten at intervals. The object of feeding the salt at watering places and from boxes is to keep them from licking the dirt where salt has stood and to keep them from eating the prairie alkali. In addition to the opportunities afforded them of being infected with parasites from the salted ground there is the injurious effect of the swallowed sand. This sand often packs away in the gall ducts and produces disease.

It may be when the gap in the life history of *Tænia fimbriata* is known that a single measure of prevention will eradicate it from the flocks. Until then the general measures prescribed above are to be recommended.

There are various minor precautionary measures to be fulfilled that will help affected sheep to live through the colder winter, and finally to render effective service in spite of the parasite. The *post-mortem* examinations have led me to expect that from 80 to 95 per cent. of each flock is infected. Now, were all of these to suffer as some of the more diseased do, sheep-ranching would be at its end. Fortunately a sheep may have a few parasites and not be seriously affected by them. This is shown by the fine, large sheep slaughtered which are passably fat and yet contain *tæniæ*. It is a frequent remark of the ranchman that if he could carry his lambs and yearlings through they would do well enough afterwards. It is these younger and growing animals that succumb soonest to the parasite. It is a rule that all young and growing animals are more seriously disturbed by the presence of parasites than older ones. Young lambs born in May or June have necessarily but a short time in which to grow before the cold season. When food is plenty, and there is no disturbance of their digestion, or other ailments, they enter winter strong enough to endure the weather without particular suffering. Interference with digestion, lack of food, or any ailments render the lambs so much the weaker, and consequently less able to endure the winter storms. The parasites interfere with digestion, and to overcome their evil effect means should be taken to supply easily digestible and fattening food, which may replace and add to that lost. Many ranchmen already feed their lambs extra grain during the fall, and have learned that not only are their losses diminished, but that the lambs become larger and stronger as well as fatter.

Formerly the ranchman depended, as many do still, entirely on the prairie for grass throughout the season. Of late years many are feeding more and more hay during the winter, and find that they profit by it in the diminished death-rate and the improved condition of the flocks in spring. This fall and winter feeding is, then, to be especially recommended as a remedial measure against losses among tape-worm infested sheep. With increased prosperity, flock-masters are adding to their shedding at the home corrals. Though the first cost seems considerable, such are the evident benefits in preventing losses during the extremely cold snaps and blizzards, that not only should they be built at the home ranches, but also at the outlying ones where now, with few exceptions, none are to be found.

The water afforded sheep, more especially lambs, should, if possible, be made warm during the coldest weather. The temperature of sheep is about 103° Fah. In giving them water which is less than 35° Fah., the heat which is necessary to raise the water to the temperature of the animal is withdrawn from other portions of the body, and digestion is often disturbed and less water is drank. Experiment has proven that animals fatten better on warm water, and were it practicable no water cooler than 60° Fah. should be offered to sheep. The maintenance of the drinking-water at this temperature for the use of the lambs and other home stock would probably repay the Western ranchman, as it certainly would the Eastern farmer. This is impracticable at some ranches, but there are many home ranches where lambs and blooded stock are kept at which the system might be pursued with advantage.

There is another possible chance of infection which there is no known means of remedying. If, as is probably the case, the *tænia* embryo passes with but little modification from sheep to sheep, there is then a certain amount of infection that may occur between the ewes and offspring when suckling, the lambs becoming infected with embryo by rubbing them from the mother in nosing around while suckling. As older sheep have the *tænia*, and as lambs become infected after being weaned, this method of infection is only one of many.

No medicinal remedies or preventives can be advised. The recommendations above are directed toward lessening the chances of infection and preserving the health of the animals.

It is hoped that the gaps in the life history of *Tænia fimbriata*, or others of our unarmed *tænia*, may yet be filled out. With a knowledge of this history, the methods of prevention would be evident to all.

The flock master should take pains to examine the sheep which die and inform himself, as nearly as possible, of the cause of death. In case of the presence of tape-worms, causing sickness, he can soon inform himself of their abundance, of the absence of other disease, and of many other things. He can soon judge whether others of his flock have them, and can more intelligently set about their treatment. A careful study of each case will then place the observant man in possession of many facts which will help him in the proper management of his flock.

The above recommendations have been written with a view of keeping the food and drink of the animal as clean as possible. Other precautions will suggest themselves to the ranchman.

Post-mortem examinations.—The tables herewith presented are the results of *post-mortem* examinations of sheep, some of which were killed expressly for the purposes of investigation. Others were examined while being slaughtered for food, either on ranches or at the shambles in Colorado Springs, Colo., while in other cases examinations were made upon dead sheep found either at or in the vicinity of various ranches. From the wide-spread distribution of the disease, the notes are such as could be taken from nearly all localities, and can not be ascribed as purely local causes. Where the observations were taken from the sheep raised in States other than Colorado, the State from which they came is given.

In addition to the presence of *Tænia fimbriata*, the occurrence of *Tænia expansa* has been noted, and also of *Tænia marginata*, which occurs in sheep in its cysticercal stage. The examinations at the shambles could not be conducted with the same accuracy while hunting for *Tænia fimbriata;* the occurrence of the *cysticerci* is, therefore, omitted in the *post-mortem* observations of June 7, 1887, to August 15, 1887, inclusive.

Table A shows that *tæniæ* occur in sheep throughout the year. It also indicates a wide-spread distribution.

A.

Date.	No. sheep examined.	No. lambs examined.	Location.	T. fim-briata oc-curred in	Cysti-cerci oc-curred in-	T. ex-pansa oc-curred in-	No. of ear-tags.
1886.							
Sept. 16	1		Colorado			1	
18	1		do		1	1	1
21	1		do	1	1		
24	3		do	3	1		
27	2		do	2	1		
30	11		do	9	9		
Oct. 1	10		do	9	6		
16	5		do	4	1		
21		1	do	1			
29	4	2	do	6	3		
31	2	1	do	2			
Nov. 2	1		do	1			
3	8		do	8			
4	5		do	3			
9	2		do	3			
10	2	1	do	3	1		
11	1		do	1	1		
15	1		do	1			
16		1	do	1			

A.—Continued.

Date.	No. sheep examined.	No. lambs examined.	Location.	T. fimbriata occurred in-	Cysticerci occurred in-	T. expansa occurred in-	No. of ear-tags.
Nov. 19		1	Colorado..	1	1	1	
21		2	...do	2		2	
22	10		...do	10	7		
Dec. 26		1	...do		1	1	105
29		1	...do	1	1		107
30		1	...do	1		1	109
1887.							
Jan. 8		1	...do	1	1		108
Mar. 14	2		...do	2			103-104
29	1		...do	1	1		101
Apr. 20	1	1	...do	1			106
28	1	1	...do	1	1		102
May 7		1	...do	1	1		110
June 7	8		Nebraska.		*8		
9	10		...do		10		
13	10		Utah		8		
25	5	12	Colorado..		5		
July 29	4	4	...do		3		
Aug. 5		4	...do		2		Y. & L.
5	20		...do		19		
July		1	...do				
Aug. 15	4	;4	...do		3		
Total	136	32		139	42	5	

* From June 7 to August 15 inclusive, *cysticerci* were found in most of the animals examined.
† No *tæniæ* found in the two lambs.
‡ No *tæniæ* found in lambs.

Experiments—November 30, 1886.—Six lambs, selected from a bunch that had been kept in an inclosure since the 15th of October, were, with two yearlings and two two-year-old wethers, crated and sent to the United States Veterinary Experimental Station at Washington. The lambs were from a collection of the runts of a large flock which had been fed on hay made from prairie grass and on coarse corn-meal and bran before November 30. They were watered from tubs, the water being drawn from a well and a pond near by. The land on which the hay was grown had not been crossed by sheep since spring, at least. The water was clear. The chances of infection from these sources were small. The corral where they were kept was a good warm shed, located amidst others, with a small adjoining yard. The dogs, of which there were two and sometimes more, had free access by jumping the hurdles; but I do not remember ever having seen one in the inclosure after the lambs were admitted. With these lambs were two old bucks and a few sheep, which were either lame or otherwise ailing. They arrived in Washington, D. C., December 4, 1886, and were afterwards placed in stalls where they could not be re-infected excepting from each other. As Table B shows that there were no adult *tæniæ* in the lambs, reinfection could only be through the four older sheep confined with them. Reinfection could not possibly have proceeded from some Eastern sheep confined with them, for these sheep, when examined, had no *T. fimbriata*. No dogs were admitted to the box-stalls where they were kept. Their food was Eastern clover and mixed corn and bran. They were furnished with well-water and salt. The adults were numbered 101 to 104; the lambs from 105 to 110.

December 9.—Two Eastern lambs, Nos. 111 and 112, were put in the pen with Nos. 101 to 110 inclusive. Nos. 111 to 118 were a number of Eastern coarse-wooled sheep, bought for experimental purposes.

December 13.—Nos. 111 and 113 were found to pass mature embryo bearing proglottides of *T. expansa.*

December 17.—Fed Nos. 105, 107, and 109 with proglottides of *T. expansa* from Nos. 111 and 113.

December 31.—Lambs Nos. 106, 103, and 110 were put in a pen with Nos. 111, 112, 113, and 115. And sheep Nos. 101 and 104 were put with Nos. 114, 116, 117, and 118. Later on some other changes were made, but as the Eastern sheep were found to contain no *tænia* when examined, these changes had no result, and could not have affected the result in other ways.

The object of arranging and re-arranging these sheep was to give possible chances of infection to the uninfected sheep.

Table B is compiled from data obtained from lambs born in 1886; from four wethers, which, with six of the lambs, were removed to the experimental station in Washington, D. C., and from a few lambs born in 1887. The sheep marked M, killed June 25, was also adult.

Table B shows that the *Tænia fimbriata* begins to appear in two or two and a half months old lambs, that they continue throughout the winter and gradually attain maturity as spring approaches. Each of the tables, A and B, shows that adult tape-worms were to be found throughout the year.

B.

Date.	No. of lamb.	Number tæniæ found.	Length of T. fimbriata.	Age in weeks.*	Time in weeks.	
					From range.	In Washington.
1886.						
Oct. 21	A	Few ...	5ᶜⁿ and under ...	23		
20	B	Many...	10ᶜᵐ and under...	24		
29	C	...do..	....do ...	24		
Nov. 9	D	...do ...		25		
10	E	3.		25		
16	F	Many...	10ᶜᵐ and under...	26		
19	G	..do ...	...do ...	27		
21	H	..do ...	...do ...	27		
25	I	..do ...		28		
Dec. 26	105	4...	2ᶜᵐ ...	32	10	4
29	107	19...	1ᶜᵐ to 4ᶜᵐ ...	33	11	4
30	109	100+..	5ᵐᵐ to 10ᶜᵐ ...	33	11	4
1887.						
Jan. 8	108	Many...	1ᶜᵐ to 7ᶜᵐ ...	34	12	6
Mar. 14	103	50+...	Immature and adults	43	22	15
29	101	Many...	....do ...	45	24	17
Apr. 20	106	25 ...	7ᶜᵐ to 15ᶜᵐ...	49	27	20
28	102	15...	Immature...	50	28	21
28	104	5.	...do...	50	28	21
May 7	110	1...	Adult...	51	29	23
Aug. 5	K	2...	2ᶜᵐ...	10-12		
5	L	2...	.. do ...	10-12		
June 25	M	Many...	2ᶜᵐ to 5ᶜᵐ— adult ...			

* The age is that of the lambs and is estimated May 15.

Nos. 101 to 104 and M were adult sheep; all others were lambs. Adult *tænia* contained embryo.

Post-mortem examinations—December 26.—No. 105 died. It contained four small *Tænia fimbriata*, the largest about 2ᶜᵐ long, and fifteen *cysticerci* of *T. marginata*, each less than 1ᶜᵐ in longest diameter.

December 29.—Killed No. 107. It contained nineteen small *T. fimbriata*, the largest about 4^{cm} in length, and twenty *cysticerci*, the largest a little over 1^{cm} in diameter.

December 30.—Killed No. 109. It contained over one hundred small *tænia*, ranging from 5^{mm} to 10^{cm} in length; also a few small *cysticerci*, apparently of same age as in 107.

January 8.—No. 108 died. It contained three specimens of *Tænia expansa ;* one of these was adult ; many small *T. fimbriata*, varying 1^{cm} to 7^{cm} long ; also six *cysticerci*, somewhat larger than the earlier found.

April 20.—Killed No. 106. It contained many *T. fimbriata*, over twenty-five in all, which were over 7^{cm} in length ; none were smaller. The duodenum and gall ducts were packed. None were adult.

May 7.—Killed No. 110. It contained one adult *T. fimbriata* and several *cysticerci*.

March 14.—Killed No. 103. It contained from fifty to sixty *tænia* from 7^{cm} to 10^{cm} in ength ; four of these were in the gall ducts and were among the largest in size. *Tæniæ* immature to adult.

March 29.—Killed No. 101. Found *tæniæ* in duodenum, gall ducts, and pancreatic ducts. The gall ducts were engorged ; the liver smaller than normal. The *tæniæ* ranged in size from 7^{cm} to 14^{cm} ; three were adult. There were three *cysticerci*.

April 28.—Killed No. 104. Found five *tæniæ* from 2^{cm} to 4^{cm} in length, but no *cysticerci*.

April 28.—Killed No. 102. Found fifteen *tæniæ*. The largest were not over 7^{cm} in length, and immature. There were two *csyticerci*.

Table C is made up from data obtained from the six lambs, Nos. 105 to 110, inclusive, transported from the prairies to Washington. It shows the comparatively slow growth of the parasite; also the abundant infection of some of the animals so long as they were exposed to infection. It also presents either the possibility of infection after they were taken from the prairie or the retention of the embryo in the rumen through a considerable, time.

C.

Date.	No.	Age in weeks.	No. of Tænia fimbriata.	Tæniæ size.	Weeks after October 15.	Weeks after December 1.	Weeks after December 31.
				Centimeter.			
Dec. 26	105	32	4	2	10	4	
29	107	33	19	1–4	11	4+	
30	109	33	100+	0.5–10	11	4+	
Jan. 8	108	34	Many	1–7	12	6	1
Apr. 20	106	49	25	7–15	27	20	16
May 7	110	51	1	Adult	29	23	18

October 15, the date on which the lambs were taken from the prairie and corralled.
December 1, the date on which the lambs were received in Washington.
December 31, the date on which the adults. Nos. 101 to 104, were removed.

Lambs K and I, Table B, show that the *tænia* was developed to a length of 2^{cm} in less than ten or twelve weeks, for the number of *tæniæ* found shows a slight infection and some time may have elapsed after the birth of the lamb before its infection.

Lambs A to I, Table B, show that in from twenty-three to twenty-eight weeks the *tænia* may develop to 8 or 10^{cm} in length, and that the infection is proportional to the time exposed. The infection is, however, a variable quantity, and no definite statements can be deduced.

As the lambs do not begin to nibble grass and drink water until some few days after their birth, the development of *tænia* in K and L probably required not over two months. Lamb A, examined October 21, about twenty-three weeks after birth, gives, when compared with K and L, an approximate rate of growth of the *tænia* of 2ᶜᵐ a month, more or less. The rate of growth must so vary at different times that no definite rate can be determined at present.

The *tæniæ* of No. 105, one of the same lot of lambs as the foregoing, were no larger after thirty-two weeks than those of K and L after ten weeks. This points to a recent infection of No. 105, *i. e.*, within ten weeks, or about the time the lambs were taken off the prairie and received into the corral. The *tæniæ* of Nos. 107, 109, 108, and 106 coincide with this; but the lambs No. 109 and 108 also point to a later infection, as many very small *tæniæ* were found in them. No. 108, which had *tæniæ* 1ᶜᵐ long six weeks after its receipt in Washington, would lead us to suspect a recent infection; but this is not necessarily the case, as the influence of the rumen of the sheep in detaining the parasite for a length of time has yet to be learned. The absence of young *tæniæ* measuring less than 7ᶜᵐ in No. 106 at sixteen weeks after its last association with an animal containing adult *tæniæ* and twenty weeks after its arrival in Washington is also of interest in that it points to infection of the lamb from the adult sheep associated with it. No. 110 shows a very slight infection, and one, judging from the age of *tæniæ*, that could have occurred in Colorado.

The six cases show a slow growth of the parasite; they also point to one of two things: That the *tæniæ* are, as embryos, retained in the rumen for some time after being swallowed, or that these *tæniæ* are continually infecting their hosts by the direct method; that is, the embryos passed by sheep, with little or no preparation, pass into other sheep and develop without the intervention of an intermediary host. So far nothing has been found to absolutely prove or disprove the latter statement. The infection, as shown by the various sizes found in these lambs and other sheep, points to a continuous infection nearly all the year. (See Tables A, B, and C.)

Nos. 106 and 110 indicate a cessation of the infection for the length of time it required the smallest (7ᶜᵐ) to attain their size. Lamb A indicates the time to be something less than twenty weeks, or at the period when they were received at Washington. Nos. 106, 107, 108, and 109, which had been confined eleven and twelve weeks, show as great infection as has been seen. This would happen with animals which were being infected in confinement, for the opportunities of infection, if the infection should prove to be direct, are greater. Various conditions, as the weather, food, water, etc., have so much to do with the problem of infection that far more data are necessary.

The fact of slow development and continuous infection are the main points brought out in this experiment. Continuous infection is naturally one of the results where *tæniæ* are constantly developing and shedding ripe proglottides laden with embryos for the infection of other hosts. Continuous infection also leads me to suspect that no intermediary host is necessary for the continuance of the life of the embryo. This proceeds from the fact that no single species of mollusk, insect, or other animal is to be found at all seasons and places necessary to suit all the conditions under which we find the host infected.

Experiment No. 2.—A lamb dropped at a slaughter-house in this city was kept with its mother in an uninfected stall.

The lamb was fed on January 10, 1887, with a large quantity of proglottides of *Tænia expansa* from No. 108. The embryos were found to be alive and moving before feeding.

On March 20 the same lamb was fed with proglottides of *T. fimbriata* from No. 103. These contained live embryos on the 18th instant. ·

On March 29 fed the same lamb with proglottides containing embryos of *T. fimbriata* from No. 101.

On April 19 this lamb was killed and nothing was found except a few white spots in and on the liver. The experiment had no results.

AN EXPERIMENT TO INFECT LAMBS WITH TÆNIA FIMBRIATA.

Experiment No. 3.—May 23, 1887, placed fifteen ewes with unborn lambs in three box-stalls. They were fed on alfalfa, hay, corn, and bran. Their water was drawn from a hydrant near by. The ewes being thin in flesh, and taken from the prairie grass and placed upon dry feed, thrived but poorly. Between May 23 and May 29 eleven lambs were born, which lived until the close of the experiment. Five of the largest and oldest were placed with their mothers in stall No. 1. The remaining were divided between two stalls, Nos. 2 and 3.

These animals were removed from all sources of infection through food and water, and the lambs had never been exposed to any source of infection. The ewes were suspected of being infected with *T. fim-briata*. If the lambs became infected they would either take them of their mothers or from their feed. Between May 26 and June 15 the lambs were fed in stall A several times each with a number of ripe proglottides from adult *tæniæ*. An interval was left between each feeding, and each lamb was fed at least three times. The other lambs were not fed. All but two of the ewes were found to contain adult *tæniæ* when examined later.

The lambs and ewes were killed in nearly equal lots on June 25, July 15, and August 1.

June 25, killed one ewe and one lamb from pen No. 1, two ewes and one lamb from No. 2, and three dry ewes from No. 3. Lambs uninfected. One ewe had *tænia* 2^{cm} long. ·

July 15, killed two ewes and two lambs from No. 1, two ewes and two lambs from No. 2. Lambs uninfected.

August 1, killed two ewes and two lambs from No. 1, and two ewes and two lambs from No. 2. Lambs uninfected.

The lambs were kept for two months, and were not infected in this time. This experiment shows that either a longer time is necessary for infection or that the embryo has to undergo some development or preparation that was not allowed and of which we are ignorant. The specimens fed were taken from slaughtered sheep, examined with a microscope, and fed by placing the proglottides which contained living embryos in the lambs' mouths and waiting until they had been swallowed.

The lambs while living with their infected mothers should have been infected, providing infection by embryos fresh from the host be possible. As this was not the case, further preparation and development of the embryos outside of the ovine host seems necessary.

TÆNIA FIMBRIATA, Diesing.

PLATE XII.

Fig. 1. Adult, natural size. From contracted alcoholic specimen.

Fig. 2. Head, edge view, ×6.

Fig. 3. Head, side view, ×6.

Fig. 4. Head, top view, ×6.

Fig. 5. Portion of segment: *a*, genital pore; *b*, cirrhus pouch; *c*, seminal apparatus; *d*, the efferent tube; *e*, the rudimentary uterine apparatus; *f*, vagina; *g*, the receptacle of the semen; *h*, fimbriæ.

Fig. 6. Portion of segment more mature than Fig. 5: *a*, genital pore; *b*, cirrhus pouch; *d*, efferent tube; *c*, the uteri.

Fig. 7. The uteri enlarged.

Fig. 7a. The uteri still further enlarged, showing the contained embryo.

Fig. 8. A half-grown tænia, showing the fimbriæ, ×2.

Fig. 9. Fragment of tænia from near head, showing the lateral excretory vessels.

Fig. 10. Terminal portion of adult, ×2: *a*, segments which have lost their contractility; *b*, separated segments.

Fig. 11. External reproductive apparatus, ×40: *a*, genital pore; *b*, cirrhus pouch; *c*, penis.

Fig. 12. Adult segment showing the symmetrical arrangement of the reproductive apparatus: *a*, *a*, genital pores; *b*, *b*, uteri; *c*, *c*, fimbriæ.

Fig. 13. Embryos as they exist in the uteri: *a*, *a*, envelopes; *b*, *b*, embryo.

Fig. 14. Embryo showing envelope and its six hooks.

Fig. 15. Youngest tænia found. Natural length indicated by lines at their sides.

All specimens except figures from 1 to 4 and 15 were drawn from fresh preparations.

PLATE XII

TAENIA FIMBRIATA.
(The Fimbriate Tapeworm.)

TÆNIA FIMBRIATA, Diesing.

PLATE XIII.—Diesing's original figures.

Fig. 1. Adult, natural size.
Fig. 2. Head, side view.
Fig. 3. Head, top view.
Fig. 4. Segments near head.
Fig. 5. Segments further removed from head.
Fig. 6. Some still more remote.
Fig. 7. From near end.

PLATE XIII

1

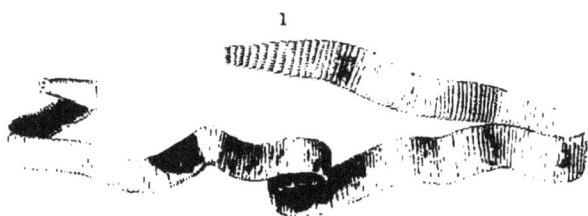

2

3

4

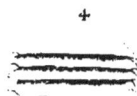

5

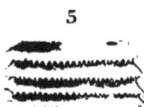

6

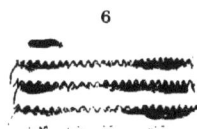

7

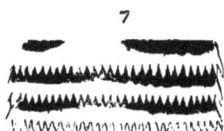

TAENIA FIMBRIATA.

THE BROAD TAPE-WORM OF SHEEP.

TÆNIA EXPANSA, Rud.

Plates XIV and XV.

Tænia expansa, the Broad Tape-worm, is one of the best known of the internal parasites of sheep, because of its flatness, length, and large size. In summer and fall it is quite abundant. The amount of pecuniary losses occasioned by its ravages depend upon the season and its abundance in affected flocks. It was introduced into this country from Europe along with the imported flocks which harbored it. Since then it has been parasitical on our flocks from generation to generation. It is now distributed from the wooded hillsides of New England to the grazing lands of Georgia, over the fertile prairies of Ohio, Illinois, Iowa, and Nebraska, and the boundless prairies, basins, and mesas of Colorado, Utah, California, and Oregon ; in short, over every sheep-grazing locality in the United States. Reported outbreaks from this vast extent of country are comparatively few and scattered, but are sufficient to warn us that when the pastures become narrowed, older, and overstocked, we may expect the same trouble with this and other parasites as has been experienced by sheep raisers through all time in the more densely populated districts of other countries.

Description.—The entire worm measures in length about 5 yards, and in width from one-twenty-fifth of an inch at the head to a half or three-quarters of an inch at the tail. Its thickness is from one-tenth to one-twelfth of an inch. These dimensions vary greatly, depending on the contracted condition of the worm when measured and on its state of preservation. Adult specimens taken from sheep may average less than 4 yards, or may slightly exceed 5 ; but they never, in this country at least, attain that gigantic measurement of 100 feet ascribed to them by European observers. The head is somewhat larger than the neck, and measures one-twenty-fifth of an inch in width. It is smooth on the end and has its four suckers directed anteriorly. Its neck, or that portion of the worm immediately succeeding the head, which is unsegmented, is short or lacking. The body of the worm is apparently made up of a series of very short but extremely wide joints, which vary in length and width in the successive portions of the body. The first rings of the *tænia* are very short and narrow; the others are longer, but are always broader than long. Those segments which are about two thirds the entire length

23038 A P——8   113 from the head are the longest. From these to the end they become gradually shorter and wider. In width the segments gradually increase from the head to the end. When first collected the texture of the worms is usually opaque and white, and it is only by allowing them to stand in water that they become transparent enough for study. It can then be seen that each segment is bilaterally symmetrical—that in each half is an independent set of genital organs. These become apparent at a little distance from the head, but show in their best development about half the entire length from the head. From this point on the segments become more and more opaque to the end. The genital organs first appear as a little horizontal line on each side ; gradually a little rosette grows at the end of each line; these disappear and the segment becomes filled with the young eggs or embryos, which form the opaque mass. The external genitals consist of a round pore on either side of each segment, in which is a minute dot, the opening of the vagina, and an exserted intromittent organ. The apparatus is called hermaphroditic, *i. e.*, each half of the segment is capable of fertilizing itself; but it is likely that cross-fertilization also occurs. Fertilization occurs about where the dots which line each side of the worm appear plainest. From this point on the eggs are developed into embryos until the end of the worm is reached, when the segments are prepared to retain vitality as individuals for some time after being ejected with excreta to the ground or water.

Besides the reproductive apparatus there are two so-called excretory canals, one on each side of the worm, running the entire length. I have not observed cross-canals, such as occur in the armed *tænia*, as in *T. marginata*.

There is no alimentary canal. The nervous system is very rudimentary, consisting of little more than bundles of nerve fibers. The suckers are each supplied with their special fibers, connected together, and each segment is supplied by two long nerve bundles which run parallel to the excretory vessels on either side. Absorption of nutrient material or feeding takes place through the surface of the segments.

The young eggs or embryos are polyhedral by pressure, and measure about 0.05 to 0.07^{mm} in diameter. They have usually two envelopes, between which a considerable amount of oily material is held. They may have three such. Around the embryo is a pear-shaped apparatus whose small end is surmounted by a cap with shredded periphery. The embryo itself is contained in a cavity in the large end of the pyriform covering, and when alive can be seen moving around in it. The embryo seems to be a highly refrangent mass of protoplasm provided with six hooks, and does not look essentially different from the embryos of other tape-worms. The cap with shredded edges is the remnant of a mass which originally included and covered the embryo and its balloon-like expansion. No attempt has been made by the writer to work out its earlier embryology.

The next stage of the embryonic *tænia* found was taken from a lamb. In this stage the young worm (about ^{2mm} long) is well outlined. It had a head with four suckers and a short unsegmented body. The next step in the development showed the body segmented. In one or two specimens I have seen a little loop with its convex end projecting towards the outside, but have not yet been able to determine its significance. Between these stages and the adult the different steps of development are easily filled in by a study of a single worm.

Occurrence and distribution.—Although there is a periodicity in the appearance of the broad tape-worms among lambs, causing at times epizootic outbreaks, the worm may be found throughout all months of the year in localities where it occurs. No section of the United States seems to be entirely free from it. They have been found in winter, in spring, in summer, and in the fall, in the intestines of lambs examined at the abattoir. They are not so frequent in winter and early spring as at other times, but seem to be more abundant in some localities than in others, though this may be due to unequal opportunities of observation in all places. A less number was found in the West than occur in the East. Conditions of climate and soil also seem to have some influence on the appearance of the tape-worms. Damp, warm climates and heavy, moist soils appear to be more favorable to their preservation while on the ground. But none of these factors would seem so favorable to the growth and life of the *tænia* as holding the sheep on over-fed pastures, as demonstrated by experiments elsewhere related.

Differential description.—*Tænia expansa* differs from *T. fimbriata* in the method of shedding its segments. Instead of maturing a few of the segments at a time and shedding them, as the latter do, whole sections ripen and pass away, so that an examination of a sheep which has been observed to pass proglottides during a past week will reveal but little more than the head of a worm. The exact length of time that is consumed by the worm in maturing, so that it all passes away, is undetermined, but it is nevertheless an important factor in the disease, for after the worm is passed the lambs begin to recover. From what was learned and seen of the disease the opinion was formed that the worms do not retain their adult size more than a month. The heads, which are left, develop slowly and form new adults. They may not cause so much disturbance at this time, however, for the lamb becomes older and better able to withstand them.

The rate of growth of the broad tape-worm is very rapid as compared with that of *Tænia fimbriata*. This is easily demonstrated by the fact that *T. expansa* are found 2 to 5 yards long in lambs from two to four months old, while *T. fimbriata* are scarcely as many inches long in lambs of the same age. If we suppose the lamb to become infected during the first month of its life, the age of the *tænia* in the above lambs would not be over three months, which would give an average growth of nearly a yard per month. This enormous increase in size is an important factor in rapidly developing disease in the young lambs. The *Tænia fimbriata*, on the contrary, slowly develops a disease which culminates in older lambs.

Life history.—The life history of *Tænia expansa* is only incomplete in that portion of its life which it passes outside of the host. Just exactly what happens to the embryo-containing egg, between the time that it escapes until it is again found in sheep as a little head with four suckers and a short tail-like appendage is not known, but from our present knowledge may be inferred with a tolerable degree of accuracy.

No one has yet been able to either develop these embryos in water or to feed them and produce an infection in sheep. So it has been supposed by reasoning from the life history of other forms of *tæniæ*, that these embryos must pass a portion of their development in some of the minute animals which inhabit the grass and water of sheep farms. From my own studies, although I have not yet been able to produce tape-worm disease by feeding the embryos, I think that the above view is fallacious, and that these embryos need not pass any of their existence in other invertebrates. Dr. F. A. Zürn (*Die Schmarotzer*, p. 191, 1882) is authority for the statement that " the disease is also present in sheep which have been fed entirely in the stalls, though more especially among the younger and youngest of a herd which are sent to the pastures."

Experiment to demonstrate method of infection.—About the middle of May, 1888, six lambs, from three to four months old, were bought on the market and added to the flock at the Experimental Station of the Bureau. This flock was kept in a small stable with an adjoining hillside yard. They were fed on clover and grain from the market, and the water was drawn from a well near at hand. The yard was sufficiently large to be grassy, but they soon ate it down to the roots. In one corner of an adjacent pen was an iron trough, kept full of water. After a rain the water might have stood in the yard for a day or two in a small puddle, but there was no so-called permanently standing water which could have harbored insect life. There were already on the place three lambs, with their mothers, which had been raised there that season.

May 16.—Two lambs were fed by drenching with the embryos or eggs of *Tænia expansa.*

May 22.—An iron trough was prepared with a grass bottom, and then filled with water. A quantity of segments of *T. expansa* were scattered in it, and at first only two of the lambs were allowed access to it. Afterwards, all were allowed to go and drink out of it.

June 11.—Slaughtered one of the lambs, which had been drenched with *T. expansa* embryos May 16, and had since been held in the yard with the trough prepared on May 22. No *tæniæ* were found. The experiment was therefore of negative value.

After these dates the lambs were neglected, so far as feeding experiments were concerned, until fall.

June 21.—One of the experimental lambs, which had previously been fed with ripe segments of *T. expansa*, was killed. It was in poor condition. No *tænia* were found in it, but numerous scars of *T. marginata* furrows on the surface of the liver. It was also found that numerous white patches, which were scattered along the mucous coat of the small intestine, were due to a species of *coccidia*. As the latter disease has not been seen since that time, and as there was a possibility that the lamb had become infected with the *coccidia* from the dirt thrown out of neighboring rabbit pens, where the disease was abundant, it has not since been studied. The results, so far as *T. expansa* is concerned, were negative.

October 10.—A ewe lamb was examined which had died some hours previously ; but one *cysticercus* was found.

October 13.—A young buck lamb was examined, which was bought in May with his mother, soon after birth, from a neighboring slaughter house and was one of those which is referred to as raised on the place). There were found a quantity of young *tæniae*, many adult and young of *Strongylus contortus*, a few young *Dochmius*, and a few *Trichocephalus*. This lamb was one of the two that was drenched with eggs on May 16. He had pastured with the others, which it was subsequently learned had adult *tænia*, and which had been purchased supposing them to be infected.

October 9.—A buck lamb was examined, one of the six purchased in May. Old cicatrices of *Tænia marginata* were found in the liver, six *Cysticerci* of this species, two adult and six young *Tænia expansa*, a quantity of young and old *Strongylus contortus*, and a few specimens of *Trichocephalus*.

October 16.—A buck lamb, another of the six, was found to be infested by a quantity of very young *tænia*, also by young and old *Strongylus contortus*, *Dochmius*, and *Trichocephalus*.

October 17.—Another of the six animals bought in May was found to contain young and adult *Tænia expansa*, *Strongylus contortus*, *Dochmius*, and *Trichocephalus*.

January 3, 1889.—A lamb slaughtered for examination was found to contain three *Tænia expansa*, one *Cysticercus*, and many specimens of *Strongylus filicollis*.

The period which had elapsed from the time that these lambs had been received on the place, to the period when the majority were examined (from May to October), was about five months. Those first killed gave negative results, probably because they were examined too soon after feeding. The lot examined in October gave very positive results. The worms found in them varied from very young to adult. The smallest and youngest are those shown on Plate XV, figs. 8, 9, and 10. These were not fed to the sheep, for lambs which had not been fed were infected ; but the majority of the young tape-worms were acquired and developed after the sheep came under the experiment. This is amply proven in the case of the young buck bought and put with its mother in the experimental yard before the former was two days old. That the time of development of the adult worm is less than four months was also proven, from the fact that other lambs, less than four months old, examined in May at the slaughter house, contained adult tape-worms. The lambs, therefore, acquired those parasites on the place. The question of the necessity of an intermediary host is also settled by this experiment, for none of those invertebrates, which are usually suggested as being the intermediary bearer, were present at any time ; nor were the conditions which are essential to the life of many such invertebrate hosts present. The pasture was and is a very dry hillside yard, from which the grass was eaten very close by the sheep.

There is another phase of the question which is still in doubt. Early

in the experiment segments of tape-worms were placed in an iron trough
out of which the sheep drank, or could have drunk all summer long;
and there might have been times when, for a day or two, a little puddle
of water could have collected after a rain. In these, more especially
the former, the development of the parasites could have proceeded
until they were taken up by the lambs. The point in doubt is, whether
the lambs got their embryos from the water or from the yard while
grazing. I am inclined to believe the latter, for they acquired other
species of worms which were not placed in the iron tank, and these were
also in various stages of development. Still another feature inclines
me to this view. If the lambs had been infected from the iron tank
they would very likely have been infected by a large number of *tæniæ*,
all of nearly equal size; but they were not. Direct infection has been
tried before, but only negative evidence obtained. The failures probably arose not only from expecting results too soon, but also from not
preparing the infecting material properly.

Since writing the above, two lambs have been examined, which give
additional data:

August 10, 1889.—A five months old lamb was examined, which had been born at
the experimental station, and kept there under the same conditions as the other
lambs, i. e., water supplied from a pump and pasturage from the yard and lane near
the sheep pens. This lamb contained two adult *Tæniæ*, and *Strongylus contortus*, *S.
filicollis*, *S. ventricosus*, *Dochmius cernuus*, *Trichocephalus affinis*, and *Œsophagostoma
Columbianum*, in all stages of growth. The last species did not show adults.

August 10.—Examined a lamb eleven weeks old, which had been bought with its
mother from a neighboring slaughter house when two days old. The lamb had been
kept under the same conditions as above. There were found one adult *Tænia* (shedding proglottides carrying well-developed six-hooked embryos) and all the other
species enumerated above, but not in the same abundance. Each of these lambs
showed that the infection had been continuous. The elder of the two yielded a
greater number of *S. filicollis* than any sheep hitherto killed, and led me to think
that this species may have been productive of more trouble than had hitherto been
suspected.

The above experiments were planned with an aim to obtain infection
within a limited area, and under conditions which could be controlled,
deeming it better to obtain infection under such conditions, though
there be a number of them to complicate the question, than to restrict
the conditions and not get an infection.

The presence of the adult *tænia* in the comparatively young lamb of
eleven weeks shortens the limit of time of complete development of the
tænia. The *tænia* was about two yards long, and had developed inside
of three months. A reason for the early infection of this lamb was that
its mother died and left it to shift for itself. These experiments con-
clude the series for determining whether sheep necessarily get the worms
from drinking water or from the pasturage.

Summary.—The life history seems from the above to be a comparatively simple affair. The embryos pass from sheep to sheep and develop
into adults, which reproduce young for infection of other animals.

Whatever changes the embryos may pass through outside the sheep can have little to do with the case as far as a knowledge of prevention of infection goes, for but very few of the conditions under which these sheep were kept can be improved by the flock-master.

Disease.—The tape-worm disease can be diagnosed by finding the little white oblong tape-worm segments which are voided from the sheep and stick to the moist pellets of dung. They may also be found adhering to the wool and dirt around the tail. But this is only after the tape-worms have become adult and have begun to shed segments. Though sheep often harbor tape-worms and give no evidence of their presence until after slaughter, there are other cases in which their presence is only too evident to the flock-master. The first indications of the disease are usually unobserved, because of the slow growth and of the comparatively small number of parasites that may be developing. The time of growth occupies about two or three months from infection. The number of individuals may be from two or three to a hundred, but it is unusual to find more than a half dozen adults together. As many as fourteen adults were found in a lamb four months old. When young they occupy but small space, need little food, and cause few vermicular contractions. In the earlier stages it is plain that they cause but little trouble, but when they grow so large that they seem to fill the whole of the small intestines they cause the serious disturbances ascribed to them. These disturbances may be to a certain extent those arising from a reflex irritation of the sympathetic and spinal nerves, but most of them seem to be due to the indigestion which the worms produce. The worms obstruct the intestinal canal by their great mass, irritate it by their vermicular contractions, cause excessive secretion of intestinal fluids, non-assimilation of food, and abstract nutriment from the intestinal contents for their own growth.

The lambs become poor and hide-bound; their flanks may either be distended by gas in the bowels or be tucked up from gauntness. In the progress of the disease the animals become evidently weaker, the mucous membranes paler, and the fleece dry and harsh from the loss of its yolk. The animals walk with a tottering gait. They often eat more and drink oftener than those less affected. In the severest cases the lambs grow extremely weak and poor, diarrhea becomes more and more pronounced, and at last they die through sheer exhaustion. While suffering from these worms they are more susceptible to the attacks of other parasites, and other diseases supervene and hasten the death of the already weakened animals.

Prognosis.—Though the tape-worm disease in its mildest form is very destructive to lambs and yearlings, yet it would seem that if they are able to pass safely through a certain period they are very apt to recover. In 1887, in the examination of two wethers which two weeks before had been passing proglottides, or segments, in abundance, and from which it was expected to secure specimens for illustration, only a small piece of the worm was found, all the rest having passed away.

Occurrence of the disease.—The worm, though present throughout the year, is more abundant in the locality of Washington during May and June than at any other season. This fact was doubtless somewhat dependent on the age of the lambs examined, which were about three or four months old at that time. In Colorado an outbreak was heard of in a flock of Merinos which occurred annually about July and August, after which time the lambs would improve. The disease is more prevalent in the summer season, and causes the greatest damage in lambs less than six months old. If the young animals can be carried beyond this age they seem to be either better able to withstand the ravages of the parasite, or to have reached a season unfavorable for its development.

Duration.—The broad tape-worms do not last long in their adult state, but after maturing nearly all their segments are shed at once. From the time that the segments are shed the afflicted lambs will begin to receive and rapidly lay on fat. The disease leaves no traces other than debility in the early stages of recovery.

Preventive treatment.—Treatment for the prevention of this disease is that suggested for general prevention of parasites and an observance of those measures which promote good health in the flock. Do not overstock pastures. Give good, pure water. When possible, put the lambs on new pastures. Feed some grain, put salt where the animals can take it daily, feed hay from racks, and grains, salt, and water from troughs.

The medical treatment promises better results than that for *Tænia fimbriata*, since the *T. expansa* is never found wedged into the gall duct or pancreatic duct, as is *T. fimbriata*, but is found lower down in the small intestine, from whence it can be removed. Many of the popular tape-worm remedies are said to be efficacious, but as the disease is difficult to diagnose until the worms begin to shed their segments much damage is done to the health of the lambs before treatment begins. On those farms and ranches where it appears periodically the lambs should be treated as soon as they begin to show symptoms. Even then complete cure can not be attained, for the lambs will continue to pick up eggs as in the first instance. For safety all sheep in the flock should be dosed, especially if all are to occupy the same pastures as those affected.

Zürn (*Die Schmarotzer*, etc., p. 191, 1882), says that treatment is practical when the disease is recognized before the lambs and yearlings are reduced to a cachectic condition. Although those far reduced in strength may not survive a medication, still they should be dosed in order that the parasites may be expelled. Otherwise the sick not treated should be yarded by themselves or killed and buried, so that they may not scatter eggs for further infection. Before giving the sheep any worm remedies they should be prepared by withholding food the night before and not watered on the morning of treatment. The dose should be administered at one time, allowing every animal to swallow it slowly if fluids are given. They should not be turned out after dosing, but should be watched during the day to see if the worms are voided. If the worms are not passed off the dose should be followed by a cathartic on the next day. If it is certain that the sheep have tape-worms and none appear, the animals should be redosed with increased quantities on the following day. Of course particular attention must be paid to the purity of the drug given.

After the sheep have been driven out the yard should be cleansed by removing the surface earth. This dirt should be placed where it can not be washed on to the grass to which the sheep have access; or, it may be thoroughly disinfected, burned, or buried. Cleansing the yard may save a reinfection.

Zürn (o. c., p. 192) details experiments made by Schwalenberg, in which wormseed, Persian insect powder, petroleum, Chabert's oil, kamala, kousso, and koussin were tried. The last three gave good results. In the first experiment 3.75 grams kamala (about 1 dram) were given to each lamb. This dose caused diarrhea and removal of the tape-worms in forty-eight hours. The lambs remained poor for a long time, in spite of good care.

In the second experiment 7.50 grams kousso (nearly 2 drams) given each lamb gave good results.

Koussin, also known as tæniin or brayerin, in 12 centigram or 2-grain doses, gave better results. The tape-worms were expelled. The treated animals remained cheerful, retained their appetites, and improved in condition.

Picric acid, 10 to 20 grains, made with meal and water into a pill, is also recommended for lambs. This quantity is sufficient for one dose, and should be followed by a cathartic. Epsom salts in 4-ounce doses is a good saline cathartic, or 4-ounce doses of the bland oils, administered slowly, may be used.

The powdered male-fern root, in 2-ounce doses, is recommended, or the ætheric oil of male-fern in dram doses. The latter is the best. It can be given in combination with from 2 to 4 ounces of castor oil. Dr. H. Pütz (Seuchen und Herde Krankheiten) recommends dosing in the morning, and withholding all food the night before giving the medicine, and on the following morning to give a cathartic. This may be unnecessary, however, when the male-fern has been given with castor oil.

Fröhner (Thierärztliche Arzneimittellehre, 1889), gives the following recipes for lambs with tape-worms: Take of koussin 3 grains, and of sugar 10 grains, mix, and give at one dose. The dose of tansy is from 2 to 6 drams. It forms one of the chief ingredients of Spinola's worm cake, which is fed to lambs as a preventive medicine against worms. The recipe, sufficient for one hundred sheep, is as follows: Take of tansy, calamus root, and tar each 2½ pounds; of cooking-salt, 1½ pounds; mix these with water and meal, make into cakes, and dry. This is an old and oft-repeated recipe, but I can not vouch for its efficiency.

Powdered areca nut may be given to lambs in from 1 to 3 dram doses. If it does not produce a stool in three or four hours it should be followed by a cathartic.

Ground pumpkin seeds are in repute with some, but it is difficult to induce sheep to eat the required dose.

Tellor (*Diseases of Live Stock*, 1879, p. 383) recommends salting liberally, and giving once a week the following saline tonic and bitter-lick as a preventive against worms: Take of common salt 2 pounds, sulphate of magnesia 1 pound, sulphate of iron and powdered gentian, each half a pound, and mix with ground feed or fodder sufficient for seventy-five to one hundred sheep. This recipe is a very satisfactory one for improving the appetite and health of the sheep, but probably can not be depended on for keeping the worms away.

For tape-worm in sheep the same author recommends: Powdered areca nut, one-half to 1 dram; oil of male-fern, 10 to 20 drops; give in molasses and water, and follow the next day with a purge. For purges he gives two recipes: Sulphate of magnesia, 2 ounces; warmwater, 1 pint in one drench; or, castor oil, 3 ounces; calomel, 12 grains; molasses, 3 ounces—for one dose.

After the sick lambs have been treated, care and attention should not be relaxed, for it is of the utmost importance that they regain their strength and vigor before the cold weather sets in. The best pasture, an extra feeding of hay, and some roots and grain in judicious quantities, should be allowed them. The feeblest should be kept by themselves, where they can get plenty of food and water without entering into competition with the stronger. This treatment should be kept up until they are fully able to hold their own.

TÆNIA EXPANSA, Rud.

PLATE XIV.

Figs. 1 and 2. Young tape-worms, natural size.

Fig. 3. Head end of tape-worm, drawn to show vermicular contractions when living.

Fig. 4. Head, top view : *a, a,* the suckers or cups, by which the worm attaches itself to the intestinal walls.

Fig. 5. Head, side view : *a, a,* suckers ; *b, b,* folds in the neck ; *c, c,* the first segments.

Fig. 6. The large end of a young tape-worm : *a, a,* segments which are not mature enough to drop off ; *b, b,* segments ready to pass away from the worm.

Fig. 7. Segments, or proglottides found, separate from the worm.

Fig. 8. An adult tape-worm, drawn in sections at regular intervals apart : *a,* head.

Fig. 9. A fragment of another worm, which is not only slightly larger, but whose segments are shorter and broader.

The specimen shown in fig. 8 could have assumed very much the same form when alive as is seen in fig. 9.

PLATE XIV

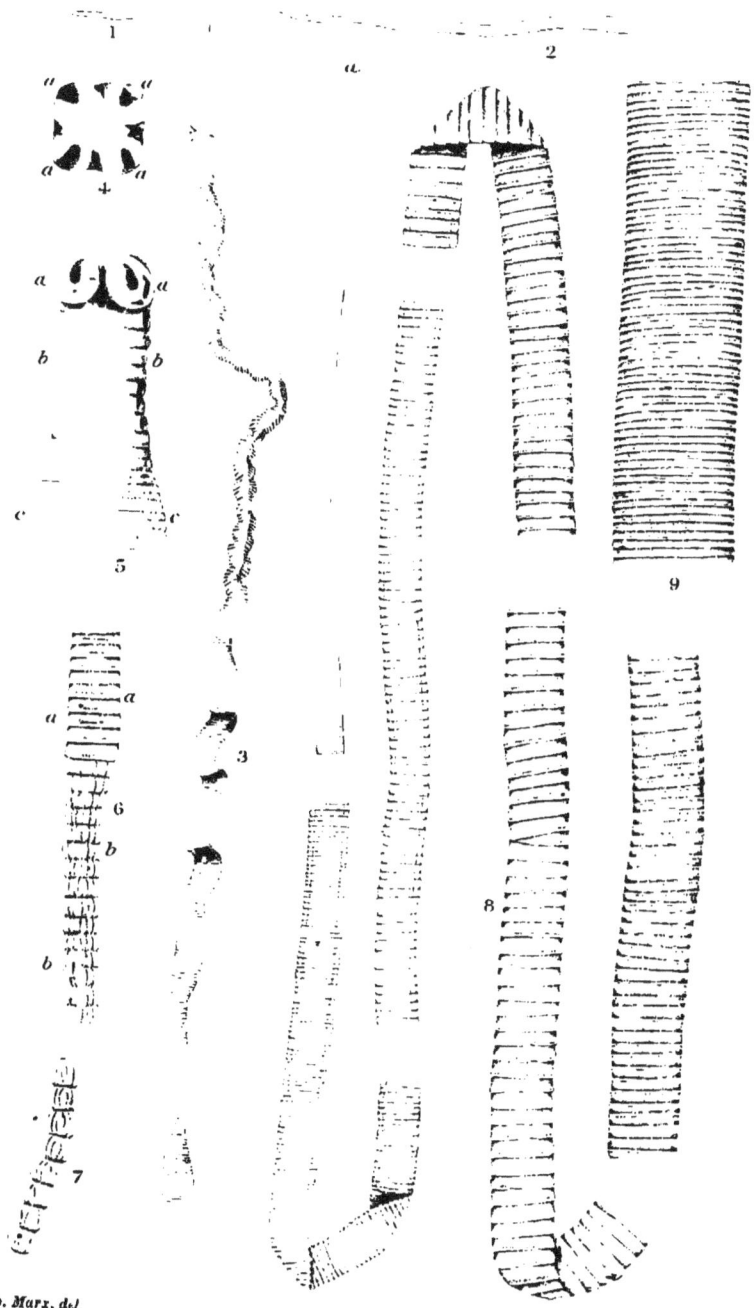

TAENIA EXPANSA,
(The Broad Tape-Worm.)

TÆNIA EXPANSA EMBRYOS.

PLATE XV.

Fig. 1. Embryos magnified and flattened under cover-glass.

Fig. 2. A single embryo and its envelopes somewhat flattened.

Fig. 3. The same greatly flattened: *a*, the thick, oily mass between the inner and outer covering,

Fig. 4. The same with outer envelope ruptured: *a*, the outer envelope; *b*, the inner; *c*, oil globules; *d*, the embryo and its pyriform apparatus.

Fig. 5. Embryo, with three envelopes.

Fig. 6. Embryo escaped from ruptured inner coat: *a*, the fringed cap-like covering; *b*, the bladder-like sac protuding from it; *c*, the six-hooked embryo.

Fig. 7. A younger embryo, in which the bladder portion has not burst from the cap; *a*, the cap; *b*, the embryo.

Fig. 8. Three young *tænia* which show no trace of segmentation.

Fig. 9. A young *tænia* which is beginning to segment.

Fig. 10. A head of young specimen showing a peculiar loop in it.

PLATE XV

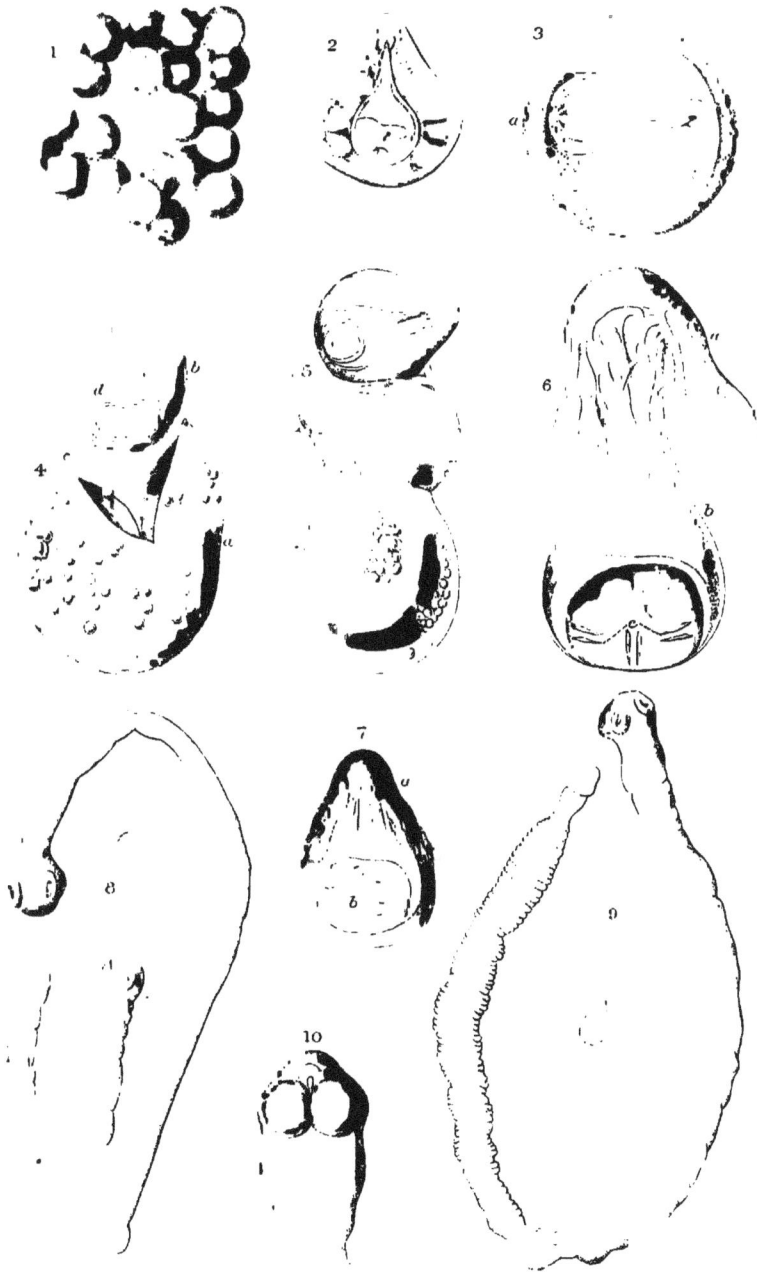

TAENIA EXPANSA
(Young Stages.)

THE LIVER FLUKE—LIVER-ROT.

DISTOMA HEPATICUM, Linn.

Plate XVI.

The liver fluke disease, which causes so much loss in Great Britain and on the European continent, is comparatively unknown in this country; so rarely, indeed, is it discovered that most authorities on the management and care of sheep seem never to have seen it. That it has occurred in this country, and that it is present in certain portions of it, is tolerably certain, for good observers have recorded it at various times.

Henry Stewart, in the *Shepherd's Manual*, 1882, page 223, says that flukes were found in a flock of Southdowns at Babylon, Long Island, and also in Cotswold, Leicester, and native sheep, presumably at the same place. In the *Tenth Census Reports of the United States*, Vol. III, flukes are said to occur in Texas and California. In the latter State they have been seen by Prof. E. C. Stearnes, of the Smithsonian Institution.

The disease occurs so infrequently in this country that the writer has seen but two cases of it, and both of these were in cattle. For a description of the malady we will therefore have to depend upon writers in those countries where it occurs more frequently than it does here.

Description.—Body flattened, leaf-like, pale brown, irregular, the adult from 18 to 31ᵐᵐ long, from 4 to 13ᵐᵐ wide, oblong, oval or lanceolate, larger and rounder in front, where it is abruptly contracted in such a way as to present a conical neck; attenuate and obtuse behind. Skin bristling with numerous little points directed backward. Oral sucker terminal, rounded. Ventral sucker large, projecting, with a triangular opening situated about 3ᵐᵐ behind the first. Intestine with two ramified branches visible through the skin and of a deep shade. Penis projecting in front of the abdominal sucker, always recurved. Vulva very small, situated at the side of the male orifice or a little behind. Eggs brown or greenish, ovoid; length from 0.130 to 0.145ᵐᵐ; width from 0.070 to 0.090ᵐᵐ. (Neumann.)

Occurrence.—This parasite has been found in the livers of sheep, goats, cattle, camels, and certain wild ruminants. It has also been found in the horse, ass, pig, elephant, rabbit, and man. It lives in the biliary ducts of the liver, and, according to Küchenmeister, feeds on blood drawn from the mucous membranes of those passages. The parasite does not appear to be equally abundant at all times in Europe, but seems to develop at various periods in sufficient numbers to cause epizootics. A number of these outbreaks, compiled by Neumann, *o. c.*, p.

463, demonstrate no periodicity or law by which future outbreaks may be suspected or predicted.

Sheep-owners and veterinarians are agreed that damp, wet seasons, and damp pastures are favoring conditions for the development of the parasite and promotion of the disease.

The life history of the parasite has been determined by Leuckart in Germany, and a little later, but apparently independently, by A. P. Thomas, in England. The former published his observations in *Zoologischer Anzeiger*, December 12, 1881, and October 9, 1882, and the latter in the *Journal of the Royal Agricultural Society of England*, Vol. XVIII, part 2, 1882, and Vol. XIX, parts 1 and 2, 1883. These authors have described most of the stages in the life history of the parasite, leaving but little to be said in addition.

The egg of the fluke passes from the biliary passages through the intestine to the ground (Plate XVI, Fig. 2a). Those that fall in favorable places develop (Fig. 2b) and finally break the little lid off the end of the shell (Fig. 2c) and escape. This happens in summer and occupies from three to six weeks. At this stage (Fig. 3) the whole body is covered with fine cilia (hairs), which enable the embryo to swim about in the water. At one end of the embryo is a little projection which can be thrust out and withdrawn, and is the apparatus by which it bores into its second host. If it does not meet one in a day or two it dies. If it meets a water snail (Fig. 6 and 6a), it is not slow to penetrate into the body, where, according to Leuckart, it lodges in the respiratory cavity. Here it encysts itself (Fig. 4), contracts into an oval mass, and rapidly grows. The name *sporocyst* has been applied to this form. The contents of this sporocyst split up into a number of bodies (Fig. 4), usually from five to eight, which develop into *rediæ* (Fig. 7). Their length at this stage is about 2^{mm}, or one twelfth of an inch. These are excluded from the sac one by one through a rent. Now each redia, in its turn, develops from fifteen to twenty *cercariæ* (Fig. 8) within it, which are evacuated in turn through an unpaired orifice situated under the neck of the redia. The cercariæ are the forms that escape from the snail, and are scattered by it in its wanderings. The cercaria, after a time of active life, loses its tail, which it has developed, and again encysts itself. (See Figs. 11 and 12 and Fig. 10.) The contents of the cyst still more resemble the future fluke, and it is at this stage that the sheep swallows it with grass. It then breaks from the cyst, arrives at the stomach and duodenum, to finally make its way into the biliary canals and grow into another adult hermaphrodite, capable of giving rise to other generations of young.

According to Thomas the encysted embryo (Fig. 4) may give rise to daughter rediæ or to cercariæ, the former to develop in the summer and the latter during the cold season.

The epitomized life history is first the egg; second, the embryo, which encysts in a snail; third, five to eight rediæ, developing from the cystic larva; fourth, fifteen to twenty cercariæ, developing from each redia and escaping from the snail; fifth, the encysted cercariæ, which after having been swallowed by a lamb develop into adult flukes.

This life history is fully as wonderful as any occuring in nature. Ac. cording to it, at least seventy-five young flukes under favorable surroundings could develop from a single egg. As the proper conditions are scarcely ever fulfilled, there are but few of those which meet with all the requirements for development.

The disease created by these peculiar parasites is known by many names, the most popular of which is Liver Rot, a term expressive of the chief pathologic symptom. Aqueous Cachexia, Rot, Rot Dropsy, Sheep-Rot, Liver-Disease, Liver Fluke, Jaundice, Yellows, Verminous Phthisis of the Liver and Ictero-Verminous Cachexia, are other names for the disease, each being expressive of some of its symptoms.

The symptoms, according to Neumann (*Maladies Parasitaires*), to whom the writer is indebted for the greater part of this article, may be divided into four periods, viz:

(1) *Period of immigration.*—This is the period in which the parasite invades the liver. In this stage the flukes are small and do not cause excessive disturbances. This period lasts from four to thirteen weeks. It is probable that it lasts as long as the weather remains favorable for the development of young flukes and sheep are kept on infested pastures.

(2) *Period of anæmia.*—This ordinarily coincides with the months of November and January, or after the lapse of thirteen weeks from infection. The sheep are less lively; the mucous membranes about the eyes, the nose, and the gums, the internal surface of the ears and the skin, are all paler than in well sheep. The appetite is still good, and the animals have a tendency to fatten, caused, according to Simonds, by a better assimilation of food from the increased flow of bile stimulated by the young flukes. This fattening stage has been taken advantage of in England by a dealer (Bakewell), who purposely exposed the lambs he wished to market early to the disease, that he might send fat lambs into the market five or six weeks ahead of his neighbors. Sometimes the appetite diminishes, thirst increases, and rumination is irregular. The conjunctiva, the mucous membranes of nose and mouth, and the skin are white, slightly tinged with yellow. There is a slight œdema (puffiness); the skin is looser, feels pasty and soft to the touch where bare; the conjunctiva is infiltrated and puffy and the eye partially closed; the wool becomes dry and brittle, is easily puffed off, and sometimes falls off of itself. Weakness appears more and more marked. There is sometimes fever and quickened breathing. Palpation and percussion indicate ascites. The droppings are normal, but contain at the end of this period numerous fluke eggs. Death may result in this stage from apoplexy.

(3) *Period of loss of flesh.*—The sheep begin to become leaner at the end of the third month after the immigration of the larvae, or about the beginning of January. The malady is then at its height. The animal becomes gradually leaner; the mucous membranes and the skin are bleached, and lose the yellow tinge. The temperature is variable and is highest in either the morning or evening. Respiration is feeble and frequent. The appetite keeps up, and the feces present nothing in particular except fluke eggs. The urine is nearly normal. The animals are listless and dejected, carry their heads low, and give way when pressed on their backs. There are frequent abortions. Nursing ewes have a clear, watery milk, very poor in nutritive elements. Their lambs are weak and thin, and usually die unless they are put to another nurse. The œdema becomes localized and increased in dependent parts. It is dispelled by walking and comes again in rest. The space under the jaws and along the trachea is a frequent seat of puffiness. This disappears during rest and comes on during feeding. It is often absent with adults in hot seasons. In the three weeks which follow the animals become still poorer despite abundant feeding, and there is generally diarrhea, œdema, jaundice, and pain on pressure over the liver. Either death occurs at this period, or the animals improve and enter the next period.

(4) *Period of emigration of the flukes.*—This is the period of convalescence and of spontaneous cure. All the symptoms grow less and finally disappear, but the cure is never complete, the changes which have taken place being irreparable. The recovery of affected sheep is exceptional.

Duration of disease.—The flukes are said by some authors to remain nine months in the liver, by others fifteen months. After this time they make their way into the intestine and are evacuated with the excrements.

Thomas says that he has seen the sickness last six years, and Neumann seems to coincide with Perroncito, that the flukes have little tendency to quit their hosts. The question of reinfection of the same animal would leave it doubtful whether such long period of infection were all due to the same parasites or to renewed generations of them.

The duration of the disease, which, as a rule, is variable, depends entirely on the degree of infection and the treatment, hygienic and medicinal, which they receive. It ordinarily lasts about six months, but exceptionally may have an extremely rapid course of a few days, death being caused by an acute inflammation of the liver, set up by the parasites, and occurring in from seven to nine days after the first appearance of the disease. Weakened by the fluke disease, the animals are more susceptible to other maladies, and these may destroy them before the primary disease has run its course.

Diagnosis.—There is little difficulty in diagnosing liver-rot in the stages at which it is first noticed if the flock-master sacrifices one of the worst affected sheep. Although he may think that it hardly pays to kill a sheep to find out what the trouble is, he will subsequently learn that a little loss in the beginning leads to a great saving in the end, and therefore becomes an investment.

In fluke disease not one or two lambs only are infected, but all the flock that have been feeding together. If the flock-master has a microscope he can detect eggs of the fluke in the droppings, but this can only be found after the fluke has matured and the disease has well advanced. Besides, it requires some skill to make the required examination. So numerous are the eggs and so characteristic is the shell with its little cap, however, that this method forms an important element in diagnosis. The droppings should be moistened with water and thoroughly picked to pieces, then spread evenly on a piece of glass and covered by another thin glass. The mass should be closely examined with the aid of a microscope magnifying from 70 to 80 diameters. This operation should be repeated a number of times if eggs are not found. The examination of the liver is the best means of diagnosis. When the gall ducts are cut open the young flukes will be found in them. They usually occur in smaller or larger nodular swellings. The structure of the liver is also characteristic, presenting a dark, soft, rotten appearance. Young flukes can be found by cutting into this organ and washing it in water. After allowing everything to settle, pour off the top and repeat the operation until the contents of the dish can be distinguished. When spread out on a flat-bottomed dish the flukes may be easily distinguished by the unaided eye, or, if very young, by the help of a small lens. If the sheep has been killed the flukes will be seen to wriggle and curl, for they die slowly in water ordinarily warm.

Prognosis.—The disease is very tenacious, and when once on a farm is difficult to extirpate. Many farms in infected localities are rendered useless for sheep raising by these parasites. When a flock becomes infected it is not to be expected that the disease will be stopped until it has caused extensive damage, and that only the most prompt and effective measures will save any of the afflicted animals.

Pathology.—The disease is seated in the liver, and all the symptoms and changes observed in other parts are directly dependent on those occurring in this most important organ. The different periods into which the various alterations have been divided are directly dependent on the periods in the life history of the flukes. For our present purpose, however, it seems advisable to give but the most brief description of these pathologic changes.

When the flukes first invade the liver they cause an inflammation, which is shown by a thickening of its mass. It becomes softer, and the surface, which should be smooth and glistening, becomes rough. These changes increase as the days go by, and the liver becomes softer and thicker. The surface becomes covered by thready fibers, as well as much rougher; the abdomen contains more or less dark-colored serous fluid; young flukes from one-tenth to one-fifth of an inch long can either be found in serum-filled cavities of the liver or attached to the outside, or perhaps free in the cavity. Still later, the liver shows places puckered up, which are caused by the reparative process and the contraction of the newly-formed tissue. There will be numerous channels beneath the serous membrane visible to the unaided eye. The biliary canals will be found thickened and enlarged at places to the size of a hazel-nut, or even larger. In these dilatations are flukes of considerable size, surrounded by a greenish black, sticky mass. Sometimes hard limy fragments are found in them. The substance of the liver becomes very rotten, and crackles under pressure by the finger. When the flukes migrate healing takes place to some degree, but the liver will never look healthy. It will grow considerably smaller and become much whiter, due to the connective tissue changes which take place.

Preventive treatment.—Thomas has formulated rules of prevention against this parasite, which are founded on its life history and seem to be eminently practical. He says :

(1) All eggs of the liver fluke must be rigorously destroyed. Manure of rotten sheep or other infected animals must not be put on wet ground. As the liver and intestines contain eggs, these, too, must be destroyed or put in the compost heap. The manure of affected animals should not be stored where there is a drainage from it to the neighboring grass. It should be mixed with lime and salt before being spread on meadows or cultivated fields.

(2.) If sheep are infected, let them be sent to the butcher at once, unless they are specially valuable and are not badly affected. If kept, they must not be put onto wet ground.

The above advice is preferable with all common stock; and the exceptions, when medicine should be tried, are very few. Stock should be very valuable to repay the expense of care and treatment in face of the large percentage of death which occurs in this disease, and the ill condition of the remnant after recovery.

(3.) Care must be taken to avoid introducing eggs of the fluke either with manure or fluked sheep, or in any other way. Rabbits and hares must not be allowed to introduce the eggs.

The most prolific source of introducing parasites is in the introduction of infected sheep, and purchasers should learn all they can of the history of the animals they buy, and not purchase from diseased flocks.

(4.) All heavy and wet ground must be thoroughly drained.

Draining is of advantage in many ways. It makes tillable land of what was before useless, either for pasturage (as shown by its infecting the flocks with a fatal disease) or for cultivation.

(5.) Dressings of lime and salt (or both) should be spread over the ground at the proper season to destroy the embryos, the cysts of the fluke, and also the snail which acts as host.

After draining, such fields should be cultivated and suitable crops raised on them.

(6.) Sheep must not be allowed to graze closely, for the more closely they graze the more fluke-germs will they pick up.

This rule is advisable wherever the sheep may pasture. Sheep on over-stocked pastures do not get enough grass, and become more infested with worms from having to eat close to the dirt.

(7.) When sheep are allowed to graze on dangerous ground they should have a daily allowance of salt and a little dry food.

Exigencies can be conceived in which sheep may be allowed to pasture on infected ground, but, in view of the fatality attending the malady, it would seem more profitable to sell the animals while they are still in good condition than to expose them further to these parasites.

Lime and salt are the two cheap preventives against this parasite. The former, mixed with manure, increases its value as a fertilizer. A solution of three fourths of 1 per cent. of salt in water has been found by experiments in the laboratory to kill fluke embryos. This weak solution might prove too weak in the field, when the salt would be still further diluted by dew or rain. Perroncito has shown that the encysted cercariæ and the larvæ encysted in *Limnæa palustris* die in 2 per cent. salt solutions in five minutes; in 1 per cent. solutions they die after twenty or thirty-five minutes; 0.65 per cent. solutions kill in about the same time; in 0.25 per cent. solutions the worms live after twenty hours' immersion.

The weakness of the brine necessary to kill the parasites in the laboratory points out that a small proportion of salt mixed with the lime would be very advisable. Salt was first advised from a knowledge of the fact that sheep never became infected with flukes on salt marshes. Stronger solutions of salt also kill the snails, which are the hosts of the embryonic fluke.

The time of year for scattering the lime or salt on the fields is the first part of June, July, and August.

Neumann advises mixing two teaspoonfuls of salt for each sheep with the ground grain they eat. Perroncito advises the addition of one-half of 1 per cent. salt to the water they drink (about a heaping tablespoonful to each gallon of water).

Although the flockmaster can never completely cure his sheep, he may, by judicious medication, so improve the condition of the convalescent animals that they will take on considerable flesh. Many vegetable astringents and tonics have been tried, with more or less profit, but it is to the mineral astringents that we look for the best results. Sulphate of iron and common salt, dissolved in the drink or mixed with the food, are among the best and cheapest.

Medicinal treatment.—Some experimenters have endeavored to kill the flukes, and have met with varying success. The remedies tried have been extract of male-fern, given with turpentine or benzine, castor oil, etc. For proportions of these remedies see remedies for *Tænia expansa.* Mojkowski, according to Neumann, has obtained encouraging results against the *distomatosis* (the fluke disease) of sheep with naphthaline, given twice a day during a week, in from 10 to 15 grain doses, alone, or mixed with powdered gentian.

The following recipes, to be used as licks, may be useful in small flocks:

Take of sulphate of iron, 2 ounces; calamus root, 1 pound; of crushed oats and roasted barley malt, of each, 20 quarts. This quantity is sufficient for one hundred sheep. Other mixed grain may be substituted for the barley-malt and oats. Or, sulphate of iron, 1 ounce, and powdered juniper berries and gentian root, of each, 1 pound. Mix with 20 quarts of grits. A lick for fifty sheep.

A very complicated but apparently good tonic remedy is known as Spinola's worm-cake, see page 121. Vieth recommends the following: Oak bark, calamus, gentian root, and juniper berries, of each 2 pounds. Pulverize and add pulverized sulphate of iron, 1 pound; pulverized cooking-salt, 10 pounds. Mix thoroughly, and give each sheep a teaspoonful every two or three days. The medicine is most easily administered mixed with meal, chops, bran, etc. Either of the vegetable ingredients of the above recipes can be omitted and substituted by some other well-known tonics, though each is thought to have its special virtues. The dietary treatment is the most valuable. Grain-feeding, mashes, strong meals, as flax seed meal cakes, or cotton-seed oil cakes, can be given in judicious quantities. The general treatment should embrace every means known to the flock-master to sustain the health of the flock.

Police sanitation.—The meat of sheep affected with liver-rot is safe to eat, but in advanced stages of the disease it is too watery, lean, and innutritious to be wholesome food, and is only an inferior article. When killed during the early stages of the disease it is more salable and nutritious. Later on, it should not be put on the market or received by buyers.

DISTOMA HEPATICUM, Linn.

PLATE XVI.

Fig. 1. Adult fluke, natural size: 1*a*, young fluke, natural size. (Raillet.)

Fig. 2. Eggs: *a*, egg with developing embryo; *b*, egg with embryo; *c*, egg-shell. (Raillet.)

Fig. 3. Ciliated and free embryo: *a*, perforating apparatus; *b*, ocular spot. (Leuckart.)

Fig. 4. Encysted embryo found in snails. (A. P. Thomas.)

Fig. 5. Diagram of digestive apparatus and nervous system: *a*, mouth sucker; *b*, pharynx; *c*, œsophagus; *d*, branches of intestine; *e*, their branchlets; *f*, nerve ganglia; *g*, ventral nerve. (Raillet.)

Fig. 6. *Limnæus truncatulus*, the principal snail which is the larval host of the fluke in Europe: *a*, natural size. (Raillet.)

Fig. 7. Redia of *Distoma hepaticum*: *a*, mouth; *b*, pharynx; *c*, digestive tube: *d*, the so-called germinative cells destined to produce cercariæ. (Leuckart.)

Fig. 8. Redia containing cercariæ: *a*, mouth; *b*, pharynx; *c*, digestive tube; *d*, *d*, cercariæ. (Leuckart.)

Fig. 9. Cercaria dissected from its cyst: *a*, anterior sucker; *b*, ventral sucker; *c*, pharynx; *d* *d*, branches of the intestine terminating in cæca. (Leuckart.)

Fig. 10. Grass stalk with three encysted young flukes, *a, a*. (A. P. Thomas.)

Fig. 11. Free-swimming cercaria just before it is about to encyst. (A. P. Thomas.)

Fig. 12. A slightly older stage than Fig. 11. (A. P. Thomas.)

Fig. 13. Genital apparatus of the liver fluke: *a*, digestive tube; *b*, ventral sucker; *c*, anterior testicle; *d*, its deferent canal; *e*, posterior testicle; *f*, its deferent canal; *g*, seminal veside; *h*, genital sinus; *i*, cirrhus pouch; *j*, ovary; *k*, oviduct; *l*, shell-gland; *m*, yolk glands; *n*, longitudinal and *o*, transverse yolk-gland canals; *p*, uterus; *q*, vagina. (Raillet.)

PLATE XVI

Haines, del.

DISTONIA HEPATICUM
(The Liver Fluke.)

A. Hoen & Co. Lith. Baltimore

DISTOMA LANCEOLATUM, Mehlis.

Plate XVII, Figs. 11 to 15.

Description.—Body semi-transparent, spotted with brown by the eggs, length from 4 to 9^{mm}, width 2.5^{mm}, lanceolate, obtuse behind, attenuate forward, and terminated by the oral sucker, which is nearly as large as the ventral. Integument smooth, intestine with two branches, not further subdivided. Penis long, generally straight. Genital orifices very close to one another. Eggs ovoid, length from 0.037 to 0.040^{mm}. (Neumann.)

The life history of this parasite seems to be as yet unknown, but it is believed to be analogous to that of *Distoma hepaticum*. The embryo (Fig. 11) differs from that of *D. hepaticum* in being globular, armed with a cephalic spur, and in being ciliated only over the anterior third of the body. Its movements are said to be slower.

Ercolani has shown that young embryos measuring 1^{mm} first show traces of a digestive system (Fig. 14), and that the digestive system is preceded by groupings of cells, which first outline the testicles, then the penis, and lastly the ovaries. Two species of snails, *Planorbis marginatus* and *Helix carthusina*, which contained cercariæ, have been pointed out, the former by Willemoes-Suhm and the latter by Piana as the intermediary hosts of the fluke.

This species of fluke never produces symptoms or lesions as grave as *D. hepaticum*, and can only occasion aqueous cachexia or dropsy. This relative harmlessness is attributed by Leuckart to its small size and the absence of spines on the skin. The two species are most often found together in the same host. By reason of its minute size, *D. lanceolatum* penetrates into the finest biliary canals, where the young *D. hepaticum* can not enter. Because the individuals of the former species escape observation on account of their small size they appear to be fewer in number. Friedberger has extracted some thousands from the biliary passages by squeezing the liver. They are also found in great numbers in the gall bladder and in the intestine at the time of their emigration. (Neumann.)

The means of prevention and treatment are the same for this as for *Distoma hepaticum*. Treatment is of more avail.

PARASITES OF THE RUMEN.

AMPHISTOMA CONICUM, Zeder.

Plate XVII, Figs. 7 to 10.

Body of a rose tint, irregular, and more or less dark. It is ovoid, narrowed in front, and swells gradually even to the posterior end; obtuse and slightly recurved on the ventral face. Length from 10 to 13^{mm}, width behind from 2 to 3^{mm}. (Neumann.)

This parasite has been found in sheep in India and Australia. The author has not observed it in this country. It occurs scattered amongst the large villi of the rumen or first stomach, attached by its posterior sucker. It is said to cause but little digestive disturbance. Figures and description of this parasite are presented, as it is a representative of the genus, and may sometime serve for generic determination of allied parasites which may be found in this country.

138

LINGUATULA TÆNIOIDES, Rudolphi.

PLATE XVII.

Fig. 1. Male, natural size. (Cobbold.)
Fig. 2. Female, natural size. (Cobbold.)
Fig. 3. Egg with contained embryo. (Leuckart.)
Fig. 4. Embryo escaped from egg-shell. (Leuckart.)
Fig. 5. Pupa nine weeks old: *a,* anus; *b,* mouth. (Leuckart.)
Fig. 6. *Linguatula denticulatum.* (Leuckart.)

AMPHISTOMA CONICUM, Rudolphi.

Fig. 7. Piece of the rumen of a cow, showing the parasite attached between the papillæ by their large posterior sucker: *a,* an isolated individual, natural size. (Raillet.)
Fig. 8. Dorsal view, ×2. (Blanchard.)
Fig. 9. Lateral view, ×2. (Blanchard.)
Fig. 10. Egg of an amphistoma, ×80. (Cobbold.)

DISTOMA LANCEOLATUM, Mehlis.

Fig. 11. Ciliated embryo. (Leuckart.)
Fig. 12. Early stage of development, in which there is no digestive apparatus (Ercolani.)
Fig. 13. Another stage: *a, a,* groups of cells representing the future testicles. (Erco lani.)
Fig. 14. Stage showing the testicles plainer, *a, a;* *b,* the rudimentary cirrhus pouch and *c,* the digestive system. (Ercolani.)
Fig. 15. Adult, natural size figure by its side: *a,* pharynx; *b,* œsophagus; *c,* branches of intestine; *d,* ventral suckers; *e, e,* testicles; *f, f,* deferent canals; *g,* cir rhus pouch; *h,* ovary; *i,* albuminous glands; *k,* uterus; *l,* vagina.

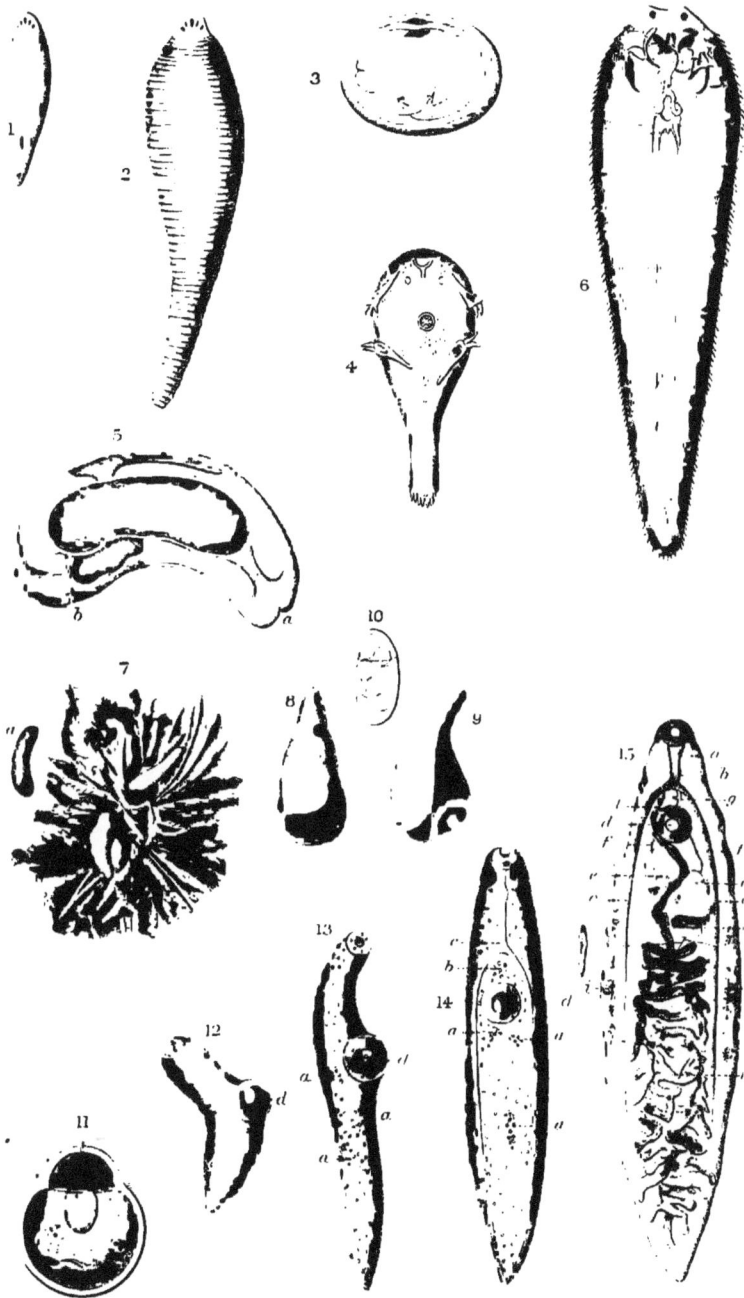

LINGUATULA TAENIOIDES.   AMPHISTOMA CONICUM.

DISTOMA LANCEOLATUM.

THE STOMACH ROUND WORMS.

STRONGYLUS CONTORTUS, Rud.

Plate XVIII.

Strongylus contortus, or the twisted strongyle, is an inhabitant of the fourth stomach of sheep and goats. Though in the majority of flocks it produces but little if any disturbance, yet there are times when, in connection with other species, it causes disease which may carry off numbers of lambs. In the southwestern States and Territories the disease has been called *lombriz*, a corruption of the Spanish word *lombrici*, meaning worms. The later writers seem to be inclined to attach a specific meaning to this word, while the Mexicans, who used it, merely intended to say that the lambs died of worm disease.

Description.—Female, 18 to 30ᵐᵐ long; male, 15 to 20ᵐᵐ long; body reddish; female marked by a double spiral white line, thicker toward the caudal end. The skin shows eighteen longitudinal lines. The mouth is round and without visible lips or papillæ. The neck has two barb-like side papillæ; unicellular glands scarcely visible. The male is about two-thirds as long as the adult female, and shows no spiral line. Bursa deeply bilobed, with a small dorsal lobe attached to one of the pair. The lobes are somewhat longer than broad. The ventral costæ are separated; the lateral are double and separated; the dorso-lateral is attached to the lateral group near its base; the dorsal costæ support the dorsal lobe and each branch is double; the twofold character is indicated by a little notch. The lateral costæ are irregularly divided, many variations being observed. Spicula two, embracing a chitinous piece between them. They are short, cylindrical, barbed on one side near the end, and have blunt tips. Female: Vulva 3ᵐᵐ from the tail and covered by a nipple-like projection, 0.5ᵐᵐ long; the latter has thin borders, and is concavo-convex, to fit the body when pressed against it. Uteri two, each opening into the common vagina; one is anteriorly directed and the other posteriorly, with a short loop between it and the vagina. The ovary of the posterior uterus is reflected anteriorly to rejoin its fellow, and together they wind spirally around the dark-colored intestine. The ovoid eggs are laid in the gastrula form, or after they have passed through the segmentation stage; length, 0.070 to 0.097ᵐᵐ; width, 0.043 to 0.054ᵐᵐ.

Occurrence.—This worm may be found in all stages in the fourth stomach or abomasum of sheep. When collected immediately after death from a slaughtered sheep they may be detected adhering by their heads to the mucous membrane. They are then of a reddish color, which may be because they feed in part upon the blood of the victim.

The life history of Strongylus contortus seems to be apparently simple. Among a number of lambs kept at the Experimental Station in 1888 were two or three which had been raised there. A *post-mortem* examination of one of these, with four other lambs which had been at the Station for the five previous months, revealed numbers of *Strongylus contortus* in all stages of growth, and of *Dochmius cernuus, Trichocephalus affinis* and *Tænia expansa*. These lambs were supplied with well water, and were allowed to run in a small, dry, grassy yard connected with a stable. The presence of these species of all sizes in the former group of lambs showed that they acquired them on the place, and that their development was direct; that is, they did not pass through a secondary host in passing between the sheep and the lambs, for all of the conditions were under inspection. The grass in the yard became very short, and probably it was because the sheep ate it so close to the ground that they became more infested with worms than sheep ordinarily do. The history, therefore, is probably as follows: The eggs fall to the ground; they are eaten by other sheep along with their feed, and they then arrive at the stomach and develop there.

The disease they cause can not easily be distinguished from that produced by other intestinal parasites. In the worst cases, besides a general lack of tone and good health, there is weakness, paleness, some fever, diarrhea, etc. In fatal cases death is said to occur within a very few days after the illness is observed; but, as the parasites develop slowly, it is probable that no symptoms of illness are apparent until after the lambs have been ailing for some time. A positive diagnosis is to be made only by a *post-mortem* examination. The little worms, if present in large numbers, will appear like masses of threads lying in the stomach. If the sheep has been killed for examination, the worms will be seen wriggling and squirming in all directions.

Treatment.—Various remedies are proposed, but of those available an emulsion of milk and turpentine, prepared by shaking the mass well, seems most practicable. Add 1 part of spirits of turpentine to 16 parts of milk, and give from 2 to 4 ounces of it to each animal, according to age of patient. One dose should be sufficient; if not, repeat it in three or four days. Or, take of linseed oil, 1 ounce; turpentine, one-half ounce, shake well and give as one dose. Quantities sufficient for any number of sheep may be made up in these proportions.

The following recipe is from Finlay Dun's Veterinary Medicine: Common salt, 3 pounds; powdered ginger and niter, half a pound each; dissolved in 3 gallons of warm water; add 24 ounces of turpentine when nearly cold. The dose for lambs between four and six months' old is 2 ounces. The entire quantity is sufficient for one hundred and sixty lambs. For delicate lambs, which are coughing and purging, the same writer recommends oil of turpentine, powdered gentian, and laudanum, 2 ounces each, all to be dissolved and stirred in 1 quart of linseed tea or lime water. This quantity is sufficient for ten or twelve doses.

Zürn recommends (after Rabe) the picrate of potash, because it is less irritating to the patients. The dose for a lamb is from 2½ to 5 grains; for an adult up to 20 grains. It can be given dissolved in water.

Dr. H. J. Detmers, in a report to the Commissioner of Agriculture, 1883, on the diseases of sheep in Texas, recommends the use of tartar emetic as follows: A half pound of tartar emetic is to be dissolved in 12 quarts of water, and from 1 to 2 ounces of the solution, containing from 5 to 10 grains of the remedy, is to be given each patient, depending on its size. He recommends dosing out of a small 2-ounce vial, and in small swallows.

Good, nourishing food, and a dry yard in summer, or a healthy, well-ventilated stable in fall and winter, are advisable. In giving medicine, drench from a horn, a spoon, or a stout glass bottle. Bottles are always liable to break. Let an assistant throw the sheep onto its haunches and hold it between his legs, back toward him. With the lower jaw seized in his left hand, from the left side, he can either seize the upper jaw or pull out the cheek-pouch with his right. The medicines are best administered while the sheep are thirsty. Small doses may be diluted, but a dose of 4 or 6 ounces is more apt to run directly into the fourth stomach than larger doses; otherwise, some of the latter might be diverted into the second stomach and fail of an immediate effect.

The following recipe was recommended to the readers of *Field and Farm*, August 7, 1889, as a preventive remedy for worms in sheep. Mr. G. B. Bothwell, of Breckenridge, Mo., who used it for fifteen years with success, is its author.

Salt, 1 bushel; air-slaked lime, 1 peck; sulphur, 1 gallon; pulverized rosin, 2 quarts; put in trough with cover, where sheep can have free access. When sheep become thoroughly infested with worms death is almost sure to follow, but the above, if kept before the sheep, will surely act as a preventive.

A more complicated arsenical recipe for worms, the source of which is unknown, is as follows:

Take of arsenic, washing soda, and carbonate of soda, each 1 ounce; put them into 2 quarts of hot water; boil, and stir for one-half hour, then add 10 quarts of cold water. The dose for a lamb, after weaning, is one-third of a gill. If the lamb is not very sick give but one dose, but if badly affected repeat in nine days.

STRONGYLUS CONTORTUS, Rud.

PLATE XVIII.

Fig. 1. Adult female, ×6: a, head; b, ovaries wound around the intestines; c, c, uteri; d, a large papilla, just in front of and covering the vulva; e, anus.

Fig. 2. Adult male, ×6.

Fig. 3. Head: a, two barb-like papillæ; b, mouth; c, œsophagus; d, intestine.

Fig. 4. Eggs, highly magnified: a, eggs before they have left the ovaries; b, eggs showing nuclei; c, eggs after they have passed through the oviduct; d, egg with one cell; e, with two; f, with four; g, with eight; h, with many; i, egg as it is laid.

Fig. 5. Skin, showing nine of the eighteen longitudinal lines.

Fig. 6. Portion of female: a, the intestine; b, b, the ends of the ovaries.

Fig. 7. Caudal end of female: a, the anus; b, the vulva; c, vagina; d, d, uteri filled with eggs; e, oviduct; f, f, ovary; g, intestine.

Fig. 8. Spicula, enlarged.

Fig. 9. Bursa, expanded to show costæ: a, ventral; b, ventro-lateral; c, lateral; d, dorso-lateral; e, dorsal; f, spicula.

Fig. 10. Group of adult males and females, natural size.

Fig. 11. Caudal end of male: a, bursa; b, spicula; c, seminal reservoir; d, intestine.

PLATE XVIII

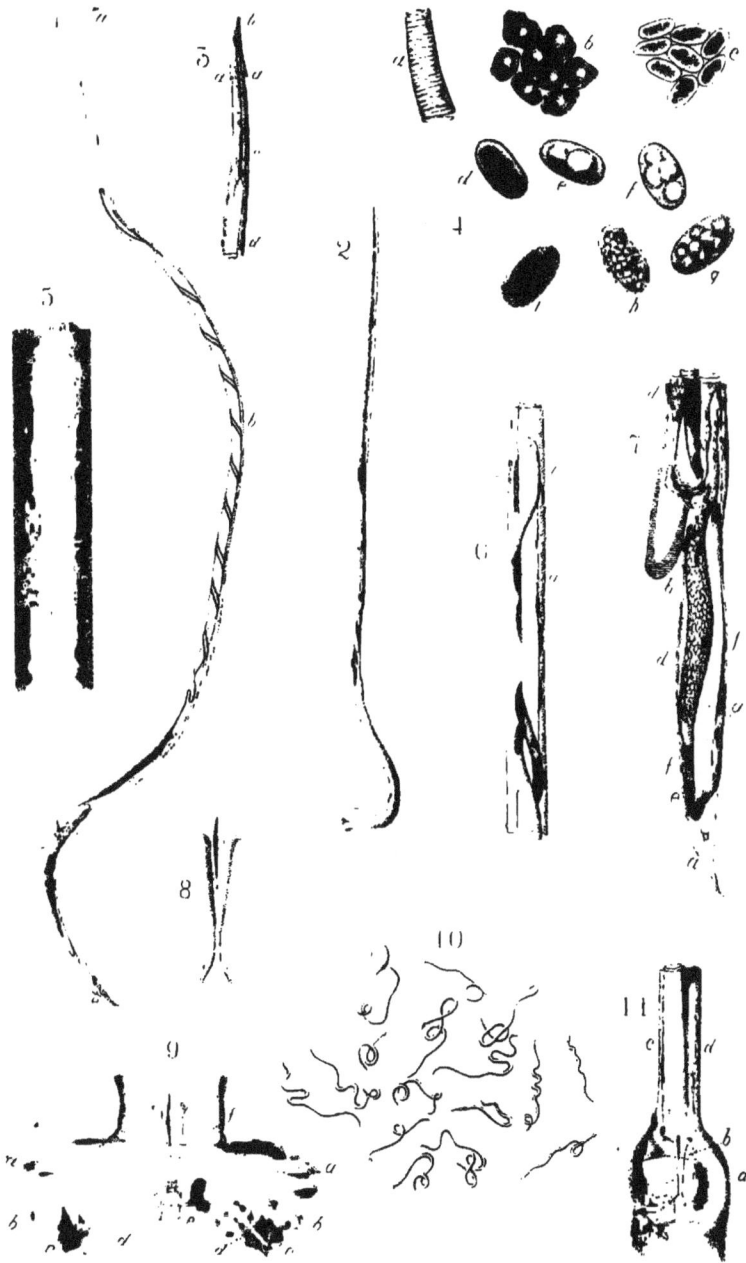

Geo. Marx, del.

STRONGYLUS CONTORTUS,
(The Twisted Stomach Worm.)

Sheep in this country harbor at least six species of round worms, parasitic in the small and large intestines, which their ancestors brought with them from Europe. They are: *Strongylus filicollis*, Rud.; *Strongylus ventricosus*, Rud., both found in the duodenum; *Dochmius cernuus*, Creplin, found in the small intestine; *Ascaris lumbricoides*, Linn., also found in the small intestine; *Trichocephalus affinis*, Rud., found in the cæcum; *Sclerostoma hypostomum*, Diesing, found in the large intestine. A seventh species, *Œsophagostoma Columbianum*, Curtice, is found in the large intestine and is probably indigenous to this country. Of all these the last species produces by far the most injury. *Dochmius cernuus* is next in importance. The injury inflicted by the others may, at times, and in conjunction with other parasites, be considerable; but disease which may be ascribed to either species alone has not yet been reported. From personal observations it is believed that the number of individuals in each sheep are usually too few to ever cause extensive loss, and that their greatest harm is from the little discomforts which they may add to those produced by the more destructive parasites.

At certain seasons of the year some of the above species are abundant, while at other times but few individuals may be found. *Strongylus filicollis* and *S. ventricosus* are usually found associated together, but are so small that they can easily be overlooked, or if found may be regarded as the young of other species. *Dochmius cernuus* is about an inch long, and being large is readily detected. *Ascaris lumbricoides* is rare, having been met with only a few times by helminthologists in any country, and in but one lot of sheep by the author. *Trichocephalus affinis* is usually found in young sheep, but is met with in comparatively small numbers. *Sclerostoma hypostomum* seems to be a rare species in the East, the single instance in which it was met with in these investigations being in examining an old sheep in Colorado. *Œsophagostoma Columbianum* seems to take the place of the last-named species in the East, and is found most abundantly in spring and summer in its adult state, although it is present throughout the year. The most favorable time for collecting most of these species has been in late fall and winter. The quantities in which the various species may be found vary with the season and the flock examined, so much so that no accurate statements of percentage of occurrence or of distribution can be made.

The symptoms which these worms produce are those of general debility and indigestion. They are caused by the irritations set up in the intestines by the worms. It is found as a rule that a weakly sheep is attacked by more than one species of parasite at a time, and, consequently, it is difficult to learn the symptoms produced by either of them acting alone. Then, too, it should be remembered that symptoms are sometimes incorrectly attributed to parasites when they really result from diseases due to entirely different causes. Even the fattest sheep harbor a few parasites, and some of them many more than one would suspect from their apparent good health.

The treatment should be directed toward keeping animals in good health and in preventing them from acquiring parasites by providing them with pure water and pastures which are not overstocked. Medicinal treatment will rarely be attempted for any single species of these parasites. A remedy which would prove effective for any one of them would d o for all. Their treatment will therefore be embraced under that for *Dochmius cernuus.*

STRONGYLUS FILICOLLIS, Rud.

Plate XIX.

Description.—Male, 8 to 15^{mm}; female, 16 to 24^{mm}. Body very small; cephalic end thread-like and tortuous; caudal end, especially of female, thick and straight. Skin marked by longitudinal lines standing at about equal distances apart. Head very small, subspherical, continuous, with a swollen cylindrical neck; the length of the inflated portion is about one fourth that of the œsophagus. Four head papillæ visible; the lateral papillæ are probably present, but can not easily be made out. Mouth terminal; apparently without chitinous armature. Œsophagus linear spatulate; unicellular gland ducts present. Position of ventral cleft not determined.

Male: Filiform and uniform in size throughout its length; bursa strongly bilobed; the membrane being well filled on the dorsum but absent on the ventrum; can not be spread without tearing; costæ generally symmetrically arranged, ventral slightly separated; ventro-lateral either joined to lateral or ventral; lateral scarcely separated; dorso-lateral joined to the dorsal, dorsal notched and with the dorso-lateral form a stem, the two pairs uniting to form the dorsal stem; the lateral costæ are the longest. Spicula 1.5^{mm} long, cylindrical, very slender and dark colored; their points are tipped with an oval inflation of the membrane and are more or less firmly attached.

Female: Tail obtuse; vulva situated about one-third of the entire length of the worm from the tail; body of the egg-bearing female enlarged in front of the vulva by the swollen and crowded uterus. Uteri directed each way from the vagina, and filled with comparatively few and large eggs in all stages of segmentation. Eggs 0.17^{mm} long, 0.08^{mm} wide, ovoid; laid in the morula or gastrula stages. Embryo not observed.

This species occurs with *Strongylus ventricosus* in the upper end of the small intestine of sheep and lambs. It is often mistaken for the young of other species, and has been identified as a variety of *Strongylus contortus.* It is needless to observe that it is specifically different from any other nematode found in sheep; a glance at the plate illustrating the species is sufficient proof of this. It is quite abundant during fall and winter. European observers seem to find it rather infrequently. The species appears to be a comparatively harmless one. It is probably the young of this species which Wedl found associated with *Tænia expansa*, and named *Trichosoma papillosum*. Wedl characterized it as having a double uterus and the mouth furnished with four papillæ. Neither of these characters is inconsistent with *Strongylus filicollis*, while the fact that Wedl's species, with a double uterus, was classified in a genus which has a single uterus and spicule indicates an error.

STRONGYLUS FILICOLLIS, Rud.

PLATE XIX.

Fig. 1. Adult male, natural length indicated by line: *a*, head ; *b*, bursa and spicula ; *c*, worm enlarged twice.

Fig. 2. Adult female, natural length indicated by line: *a*, head ; *b*, vulva ; *c*, anus ; *d*, worm enlarged twice.

Fig. 3. Head: *a*, mouth surrounded by four papillæ ; *b*, œsophagus ; *c*, inflated skin surrounding head and neck.

Fig. 4. Skin: shows nine of the eighteen longitudinal lines.

Fig. 5. Cephalic end: *a*, the head ; *b*, the œsophagus ; *c, c*, the unicellular gland ducts.

Fig. 6. Bursa: *a*, the spicula ; *b*, their knobbed tips.

Fig. 7. Portion of the spicula enlarged.

Fig. 8. Ovary of female with inclosed eggs showing segmentation.

Fig. 9. Bursa spread out: *a*, ventral costæ ; *b*, ventro-lateral ; *c*, lateral ; *d*, dorso-lateral ; *e*, dorsal.

PLATE XIX

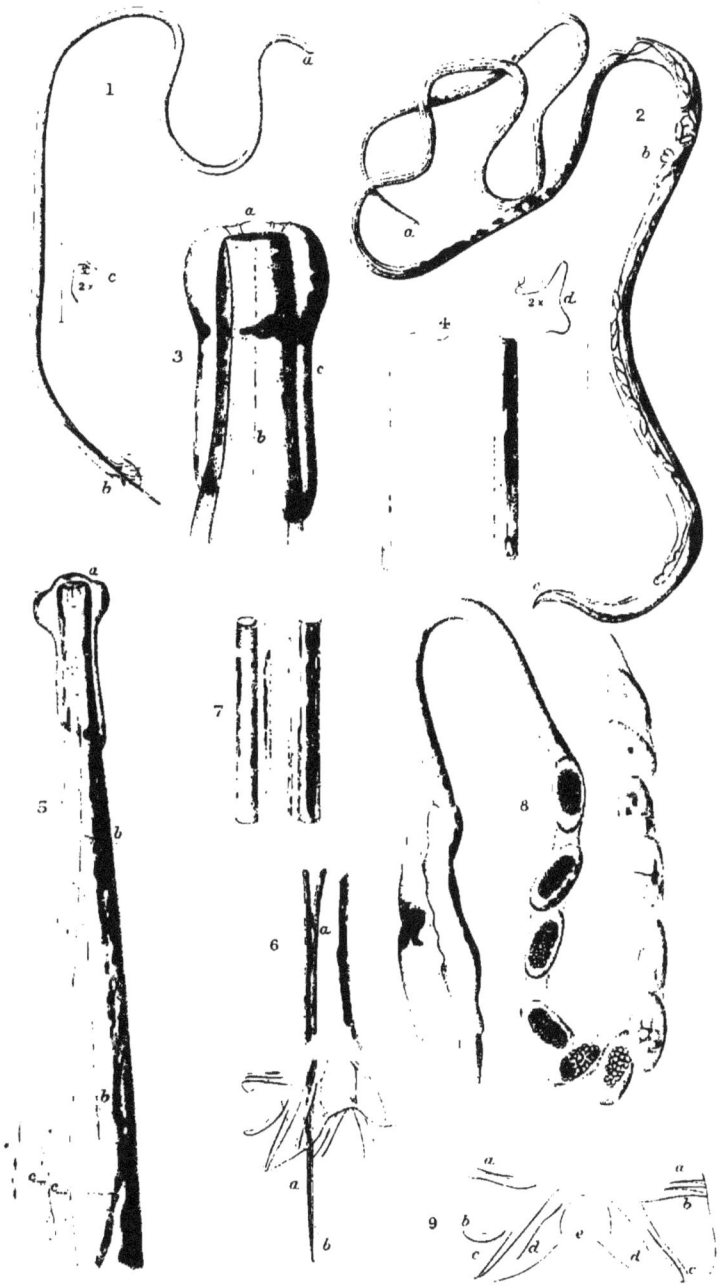

STRONGYLUS FILICOLLIS
(The Thread-Necked Worm.)

A. Hoen & Co. Lith. Baltimore

STRONGYLUS VENTRICOSUS, Rud.

Plate XX.

Description.—Male, 6^{mm}; female, 13^{mm}. Body very small and comparatively stout. Males and young females usually spirally coiled; body of old female straight, with cephalic end coiled. Skin transversely striate, marked by fourteen longitudinal lines; the five larger standing at equal interspaces on the dorsal and ventral surfaces, the two smaller standing close together on the sides. The crossings of the striæ and longitudinal lines make pits which are quite characteristic. Head little larger than neck, but hemispherical and continuous with the cylindrical inflation of the neck. No head or neck papillæ visible. Mouth terminal, very small and round. The end of the head is furnished with a hemispherical cap-shaped chitinous piece. Other oral armature apparently absent. Inflated portion of head about one-fifth the length of œsophagus. Unicellular glands not apparent. Male about one-half the length of female; bursa conical and bilobed, the ventral membrane being narrow, the dorsal wide; ventral costæ not separate, smaller than the ventro-lateral, which is stout; lateral widely separate and apparently formed of three nearly equal costæ; dorso-lateral slender; dorsal notched at the end and giving off laterally a very short side branch. Spicula 0.36^{mm} long, short and stout, and margined by a fringe-bearing sinuous membrane. They are tipped by a soft pad-like expansion of the membrane. Female characterized by a swelling at the vulva, which gives the species its name. This character is more pronounced in older specimens. Vulva from two-ninths to one-third of the entire length of the female from the tail. Uteri directed anteriorly and posteriorly from the vagina. Ova 0.13^{mm} long, 0.07^{mm} wide, comparatively large, and found in all stages of segmentation. Embryo not observed.

This species is found in association with *Strongylus filicollis* in the upper part of the small intestine of sheep. It can be separated by its smaller size, its spiral twist, and the markings of the skin. It is best found in fall and winter. The species was originally described from specimens taken from cattle, and so far as known has never before been noticed as being found in sheep. The female in its adult stage resembles the small specimens of *S. filicollis*. It is apparently the cause of little or no disturbance, although from the appearance of its mouth parts it might seem to be more injurious than *S. filicollis*.

PLATE XX.

Fig. 1. Male, natural size indicated by line and small figure of worm: *a*, head ; *b*, bursa and spicula.

Fig. 2. Female, natural size indicated by line and small figure of worm: *a*, head ; *b*, genital orifice.

Fig. 3. Head: *a*, mouth; *b b*, chitinous cap surrounding it; *c*, œsophagus.

Fig. 4. Portion of skin showing eight of the fourteen lines: *a a*, the two lateral lines; *b b*, the dorsal and ventral lines. The dots indicate depressions where the longitudinal and the latitudinal lines cross.

Fig. 5. Cephalic end: *a*, head ; *b*, œsophagus.

Fig. 6. Spiculum: *a*, its chitinous portion; *b*, the protractor muscle; *c*, the fringe edging the membranous portion ; *d*, the distal end covered by membrane.

Fig. 7. Female, natural size indicated by line : This female is more characteristic of the species than that of Fig. 2, but is not quite mature.

Fig. 8. Bursa with spicula.

Fig. 9. Bursa spread out: *a*, ventral costæ; *b*, ventro-lateral; *c*, lateral, of which there seems to be an accessory branch; *d*, dorsal-lateral; *e*, dorsal, which also has an accessory branch.

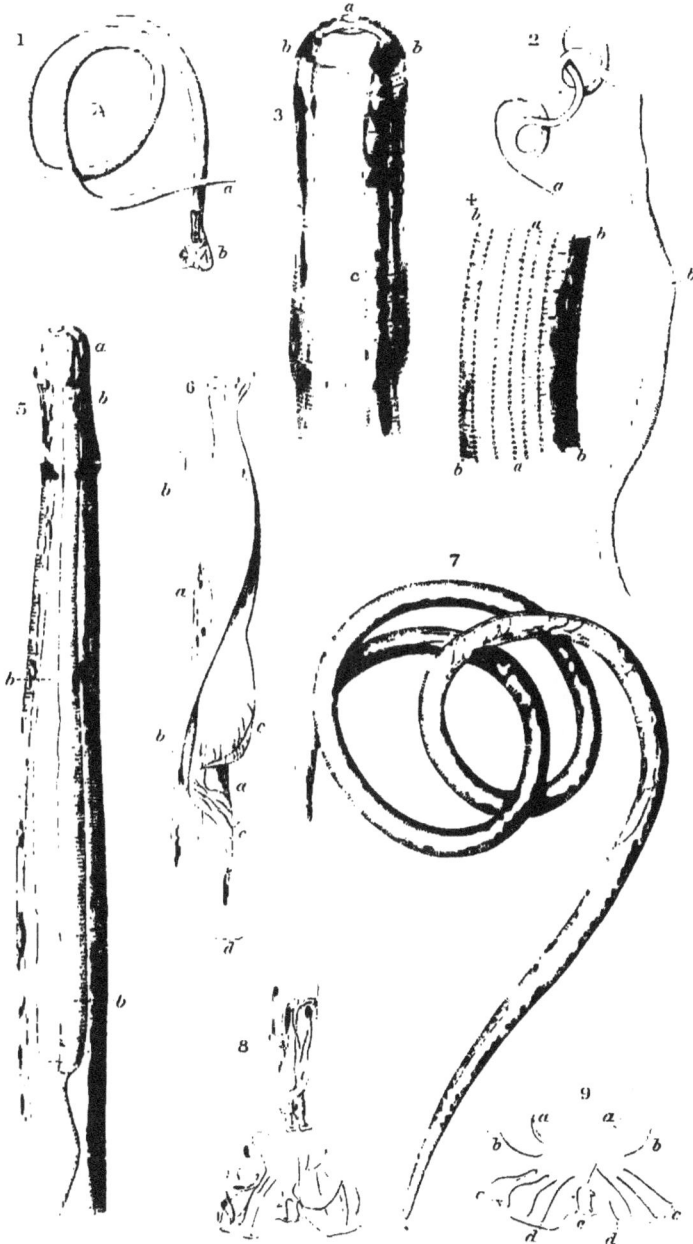

PLATE XX

A Hoen & Co. Lith Baltimore

STRONGYLUS VENTRICOSUS
(The Ventricose Worm.)

THE LARGE ROUND WORM.

ASCARIS LUMBRICOIDES, Linn.

Plate XXI.

Description.—Male and female 150^{mm} each; in the specimens figured 120 each. Body very large and thick, obtuse at each end, of a yellowish color, skin marked by rings. The site of the vulva is marked by a smooth wide band about one-third the entire length from the head. There are three longitudinal bands; the two wider are nearly lateral; the third is ventral. The head end is abruptly terminated in three well-formed lips; the dorsal possesses two papillæ near its base and the two ventral one each. Each lip consists of a chitinous support covered by cuticular membrane. The tail of each is obtuse. The description after Schneider (*Monographie d. Nematoden*, p. 36) is as follows:

"Lips nearly equal, their form changing from semi-circular to quadrangular. Teeth very fine. The lobes undivided. The azygos lobe has a rounded point, and reaches with its anterior end to the front of the saddle. Cuticular rings longitudinally ribbed. Vulva 40 to 65^{mm} from the cephalic end. Vagina 11^{mm} long. Tail of the male flat on the ventral side. Only behind the anus is the skin broadened into a bursa; sixty-nine to seventy-five papillæ on each side; the first seven pairs of these papillæ stand behind the anus; the second pair stand nearer the ventral line than the first and third; the fourth and fifth pairs and the sixth and seventh pairs are united into double papillæ. (See Fig. 7.) The succeeding stand first in a single row, then in pairs forming a double row; and on the whole they are very irregular. An unpaired papilla stands in front of the anus. Spicula two, each a single tube, with its point terminating bluntly and irregularly in front.

"This parasite occurs in the small intestine in man and swine, and sometimes forces its way into the gall ducts, stomach, œsophagus, nose, and lungs."

In one flock of sheep only has the author found this parasite. In this flock it occurred in six animals, a majority of those examined. That it is not a common parasite is attested by the writings of various European helminthologists. A few of the latter have described a special species (*Ascaris ovis*), from the sheep, but as it is infrequently found it seems to be an inconstant parasite or an adventitious one, *i. e.*, it is probably a constant parasite of some other domesticated animal, and occasionally only becomes parasitic on sheep. Besides, the specimens figured always seem to have been immature, and do not differ from *Ascaris lumbricoides* of the same size and age.

In the single flock in which this worm was observed all the specimens found were·immature. The vegetative organs of these specimens differ in no essential point from those of *Ascaris lumbricoides* as figured by various authorities. In addition, a close comparison of these specimens

151

with equally immature *Ascaris* taken from pigs gave no determinable differences. These remarks presuppose that *A. lumbricoides* of man and *A. suilla* of swine are identical species. The description seems to be incomplete, but as the species is neither peculiar to nor commonly found in sheep no attempt is made here to enlarge it. The species can easily be determined by a comparison with the figures.

Sheep probably acquire this parasite while pasturing after swine. As few are infected, little harm is accomplished by the *Ascaris*. It is more abundant in summer and fall than at other seasons.

ASCARIS LUMBRICOIDES, Linn.

PLATE XXI.

Fig. 1. Adult male, natural size.
Fig. 2. Adult female, natural size.
Fig. 3. Top view of head enlarged; a, dorsal lip with two papillæ; b, b, ventral lips with one papillæ each.
Fig. 4. Head, ventral view.
Fig. 5. Head, dorsal view.
Fig. 6. The lips greatly enlarged and flattened: a, a, papillæ; b, b, the serrated edge of chitinous support.
Fig. 7. Tail of male, ventral view: a, anus; 1, 2, 3, 4 and 5, 6 and 7, the post-anal papillæ; 8, the unpaired pre-anal; 9, 9, 9, other papillæ.
Fig. 8. The two spatulate spicula: 8a, one of these enlarged.
Fig. 9. Portion of female: a, the genital opening; b, the ventral line; c, one of the two lateral lines.
Fig. 10. Tail of female, lateral view.
Fig. 11. The same, ventral view.

PLATE XXI

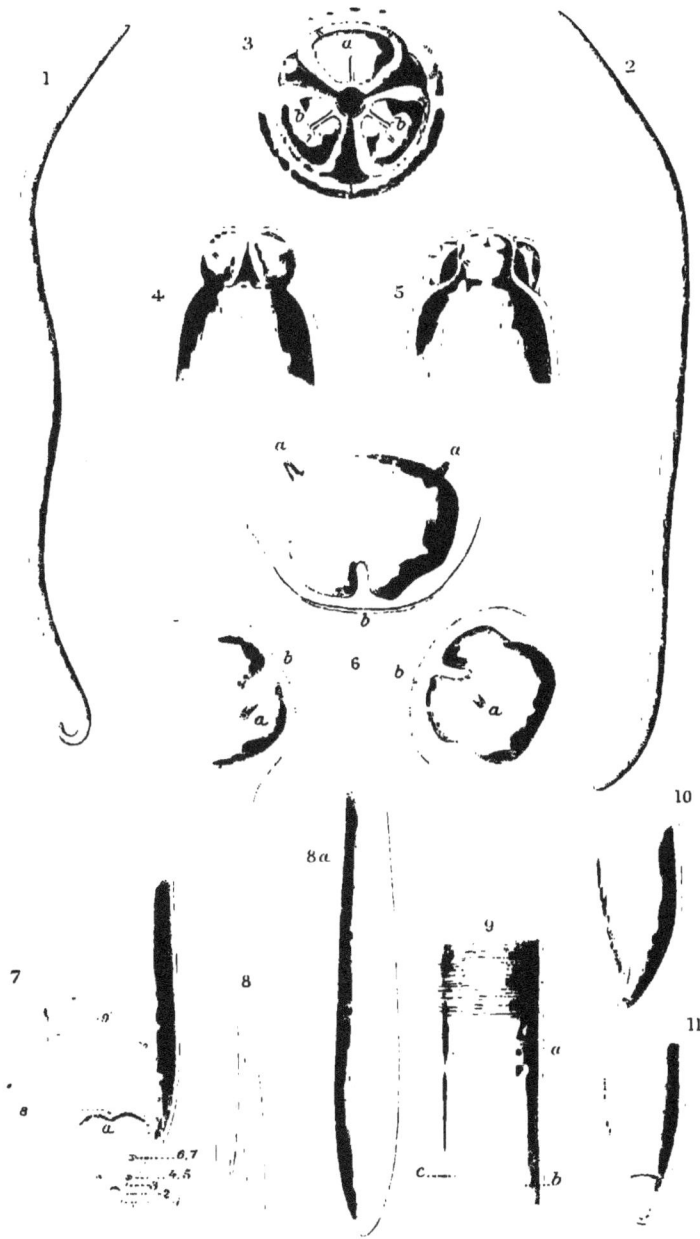

A. Hoen & Co. Lith. Baltimore

ASCARIS LUMBRICOIDES,
(The Common Round Worm.)

Dochmius cernuus, Creplin.

Plate XXII.

Description.—Female, 20 to 26ᵐᵐ; male, 13 to 17ᵐᵐ. Body very dark colored when fresh; whitened when preserved; attenuate towards the ends. Head curved, the mouth being directed dorsally. Mouth round, smaller than the oral surface, and opens into an ovoid dark-colored capsule. About the opening stand four teeth, two on each side, their base being sunk into the capsule and their free edges projecting into the cavity. The ventral are the larger, thicker, and more opaque; their edges form an unbroken sinuous line. At the caudal opening of the capsule are one pair of ventral and a single dorsal tooth; the latter is conical, very large, and rises to near the mouth. In the pharyngeal opening are six very small club-shaped, spinous, chitinous appendages of the œsophageal supports or rods; they seem to be jointed. There are six papillæ—dorsal, lateral, and ventral pairs. The lateral may give off a branch on the dorsal side. There are two lateral neck papillæ, nearly opposite the middle of the œsophagus; the ventral cleft is situated a little anterior to a line connecting the latter. Unicellular neck glands quite plain.

Male: Bursa funnel-shaped; will not spread without tearing; costæ unsymmetrical as to form; ventral pair not separated; lateral, widely separated; dorsal notched; dorso-lateral unequal in length and differently attached to the dorsal stem. Spicula 0.6ᵐᵐ long, aculeate, fenestrated, and provided with a narrow membranous margin.

Female: Vulva about three-fifths of the entire length of the body from the end of the tail. Vagina opening at right angles to the side of the body. Uteri, lying one anterior, the other posterior; each forms an **S**-like loop; the anterior ovary is directed towards the tail, and, with the posterior, forms an intricate sinuous net-work surrounding the intestine. Eggs elliptical; laid in the morula stage; length, 0.06ᵐᵐ; width, 0.03ᵐᵐ.

· *Occurrence.*—This species inhabits the small intestine of sheep, and attaches itself to the intestinal walls by its stout oral armature—the so-called teeth. It lives upon the blood of its victims.

The life history of this species of *Dochmius* has not been determined, but there is no reason for supposing it different from that of *D. trigonocephalus*, the allied species found in dogs. This life history has been determined by Leuckart (*Die Menschlichen Parasiten*, Band II, pp. 132–134), and is essentially this: The eggs pass from the dog to the ground, where, in wet places, they undergo a development of the vegetative organs. If at this stage the young are swallowed by another dog they develop into adults. The development outside the dog may consume from three to six days. The worm may then continue living without further development for an indefinite time, depending on the conditions by which it is surrounded and the favorable opportunities for being eaten by the dog. Its development in the dog occupies about two weeks. The time consumed may be supposed to be that occupied by the development of *Dochmius cernuus* with approximate certainty. Leuckart states that though he saw some of the embryos enter snails while in their free living state, that this condition was an unnecessary one, and that the worms underwent no development while in the snails. It may be that if these parasites can enter the snails, their opportuni.

ties for safely passing the indefinite time prior to finding their way into their final host are increased.

In my own experiments in keeping a number of lambs in a circumscribed space for five or six months after purchase, and in confining two others raised there with them, allowing them no water save such as was pumped for them, *Dochmius cernuus* were found of various sizes in the lambs of each set. The two lambs raised on the place must necessarily have acquired them there. These parasites either developed to a certain extent in the iron watering trough or in little pools which could have collected and remained in the yard for a day or two after a rain, or the lambs were infected from the dry hill-side of the inclosure.

The disease can only be diagnosed by the flock-master from a *post-mortem* examination. It has been recommended to diagnose these parasitic diseases from the eggs of the worm found in examining the feces by the aid of a compound microscope. Such a plan is very tedious in its execution, and impractical save to one already skilled in the work.

The disease caused by *Dochmius cernuus* receives little attention in veterinary works. This 'is due, in all likelihood, to the fact that not more than two or three hundred of the parasites ever seem to be present in one sheep, and generally there are less than one hundred; then, too, if other parasites are found present the illness would probably be ascribed to them. If, however, we may be allowed to infer the effects which would be caused in sheep from the effects which *Dochmius duodenalis*, a related species, produces in man,* in whom it has caused epidemics characterized by progressive anæmia, and if we may accumulate corroborative evidence from the disease which *Dochmius trigonocephalus*, a third species, produces in dogs, we may fairly infer that the species causes more disease than has been suspected. Nor is its comparative paucity in individuals any contra-evidence, for in human patients affected with this disease the species is represented by usually less than a hundred specimens, although as high as two or three hundred have been found in one patient. In dogs the author has found about the same number.

The intestinal lesions are obscure to the unaided eye, except at those points where the parasites have been attached. Here, if the worm has recently loosened its hold, there is a slight blood extravasation. The parasites maintain their hold by the chitinous cup with its projecting oral teeth, and in some way cause a hemorrhage, upon which they feed. The six pharyngeal spinose appendages may aid in wounding the delicate epithelial cells.

It seems impossible that a dozen or twenty, or even fifty, specimens of *Dochmius* could, by creating such little injuries in withdrawing blood

*Literature: Wilhelm Schulthess, *Zeitschrift f. wiss. Zoologie*, XXXVII, 1882, pp. 163–217. J. Ernst, *Deutsche med. Wochenschrift*, 1888, p. 291. A. Fränkel, *ibidem*, 1885, p. 443. O. Leichtenstern, *ibidem*, 1885, pp. 484, 486, 501, 523; 1886, pp. 173, 176, 194, 216; 1887, 565 and five following numbers; 1888, p. 843.

from its host, cause the severe disease and progressive anæmia ascribed to it in man, but such has been determined to be the fact from clinical and *post-mortem* observations. It may be that there is a reflex, sympathetic action stimulated by them of which we can take no account. The further changes observed in patients affected with *Dochmius* are much the same as in those affected with other parasites, except that anæmia with its attendant effects seems to be the most prominent.

The disease is one which begins in early lambhood and progressively continues, the severity depending on the number of parasites entering the intestinal canal. The adult probably lasts through the winter and continually lays eggs which pass to the ground. The character of the season, of pasturage, and of the water, in being either favorable or unfavorable to the preservation of the young worms while on the ground, will therefore determine the amount of infection and sickness during the following season.

The preventive treatment for the intestinal worms is the same as that advised for the lung worms—good care, pure water, plenty of grass, sufficient grain feeding, salt, and separation of sick from the well. For the reason that the disease has not received the attention its importance demands the medicinal treatment has not been worked out with the thoroughness that some of the other parasitic diseases have received.

Medical treatment.—In man the most effective remedy is extract of male fern, combined with powdered male fern, the remedy to be preceded five or six hours by a dose of castor oil. This combination is also a good one to administer to dogs in the following proportions: Extract male fern, 40 grains; powdered root of male fern, 75 grains. This mass must be made into ten pills with yellow wax, and all given at once. The dose of powder of male-fern root for sheep is from 1½ to 3 ounces, and of the extract from 2 to 4 drams. As boluses are not only inconvenient to give to sheep, and do not pass directly into the fourth stomach, the administration should be by drenching. I should advise that the extract be mixed with from 2 to 4 ounces of castor oil. Other remedies advised for the round worms are wormseed, wormwood, and santonine. The latter is an alkaloid obtained from a species of Artemisia. As the prairie lambs love to eat sage, of which there are a number of species belonging to the genus artemisia, it is likely that these plants may prove beneficial to them through medicinal qualities. In my examinations of Western prairie sheep I do not now recollect having met with as many round worms as are found in Eastern sheep. The dose of santonine for sheep is from 1 to 3 grains, given in from 2 to 4 ounces castor oil. The preparations of tansy, *Tanacetum vulgare*, have long been used as vermicides. The dose of the oil is from 1 to 2 drams, given diluted by adding from 4 to 8 ounces of another oil. The receipt for Spinola's worm cake sufficient for one hundred sheep is: Take of tansy root, calamus root, and tar, of each 2½ pounds, of common salt 1½ pounds, make into cakes with meal and water, and dry. The dose of areca-nut powder, which is an effective remedy for round worms as well as tape-worms, is from 1 to 3 drams for lambs.

The oil of turpentine has proven of itself a valuable anthelmintic, but should be used with care. For intestinal round worms in sheep from 1 to 4 fluid drams may be given, according to age. The turpentine should be mixed with from 2 to 4 ounces of castor oil. Sweet or linseed oil may be substituted, but their cathartic effects are untrustworthy. More than one dose should not be given until two or three days have elapsed, when, if it is deemed advisable to give a second dose, no untoward results having been noticed from the first, the dose, slightly increased, may be repeated. Do not give more than the maximum dose. Tellor (Diseases of Live Stock) gives the following recipe: Linseed oil 2 ounces, oil of turpentine one-half ounce, for a drench. The French veterinarians advise, among other remedies, the use of empyreumatic oil, petroleum, and chimney soot. Empyreumatic oil is animal oil, a by-product of distillation of animal matter for ammonia. Dippel's ethereal extract is a refined product, and the oil of Chabert consists of animal oil 1 part, turpentine 3 parts. The medicine should be administered as a liquid. The dose of animal oil, or oil of Chabert, is from 1 to 2 drams, to be given in 4 ounces of the chosen vehicle. The oils and alcohol dissolve animal oil, but if the vehicle is a watery mixture it must be well shaken. Bitter vermifuges made up into tea are excellent, as is also a decoction of chimney soot thickened by dextrine or the yellow of eggs. From the certain effect that petroleum has on insects externally we may infer that it should prove a valuable anthelmintic internally, if it may be given in sufficiently large doses. It has been used in man for *tænia* and round worms. The dose is 30 minims. The dose for sheep may be as large; how much larger future experiments will determine. Until the toxic dose is learned it should be given with caution. It is probable that 2 dram doses may be used. Give with from 2 to 4 ounces of sweet, linseed, or castor oil.

Besides these there are many other remedies proposed, as savin, sabadilla, spigelia or pink root, aloes, tartar emetic, asafetida, azedarach, kousso, kamala, and pumpkin seed. The greater part are of doubtful efficacy; others are dear, and can be replaced by remedies equally as good and cheaper. Many of the bitter herbs may be powdered and given with grain, but the sheep will not get enough to have the best effects. Often the worst-affected lamb will not eat any, or very little, on account of loss of appetite. This method of administration is, besides, wasteful. The uncertainty of sheep receiving a full dose is the chief argument against such a method. The effect of some of these plants on sheep is also modified by the fact that sheep are plant eaters and become more or less accustomed to the various medicinal principles found in them.

DOCHMIUS CERNUUS, Creplin.

Plate XXII.

Fig. 1. Adult male, × 9 : 1a, natural size.
Fig. 2. Adult female, × 9 : 2a, natural size ; b, vulva.
Fig. 3. Cephalic end of male, ventral view : a, head ; b, b, œsophagus ; c, c, neck papillæ ; d, opposite ventral cleft out of which the unicellular gland ducts e, c, empty by a common tube, f ; g, unicellular glands ; h, h, two ends of the seminal tubes ; i, i, i, seminal tube ; k, k, intestine.
Fig. 4. Head, dorsal view : a, mouth ; b, b, the chitinous oral teeth on the left side ; c, c, c, three papillæ on right side ; d, dorsal teeth ; e, e, neck glands ; f, œsophagus.
Fig. 5. Head, right side. Letters as in Fig. 4 : g, ventral teeth ; h, chitinous capsule.
Fig. 6. Pharynx and lower part of capsule of head, ventral view : a, the cut edge of capsule ; b, the dorsal tooth ; c, c, ventral teeth ; d, pharynx, in which are six club-shaped bodies, which seem to be appendages of the chitinous rods of the œsophagus ; e, œsophagus ; f, enlarged club-shaped body, which is chitinous and covered with rough points. This apparatus seems to be for mastication.
Fig. 7. Middle portion of female, × 30 : a, vulva ; b, vagina ; c, c, portions of cephalic uterus ; d, d, caudal uterus with eggs ; e, oviduct ; f, f, cephalic ovary ; g, g, caudal ovary ; h, h, intestine.
Fig. 8. Eggs : a, a, from ovary ; b, in oviduct ; the rest from the uterus in various stages of cell division ; c, as they are expelled.
Fig. 9. Bursa : a, ventral costæ ; b, ventro-lateral ; c, lateral ; d, latero-dorsal ; e, dorsal. The two sides are unsymmetrical.
Fig. 10. Spicula.
Fig. 11. Portion of spiculum showing fenestrations and openings into the tube : a, tube ; b, wing.

PLATE XXII

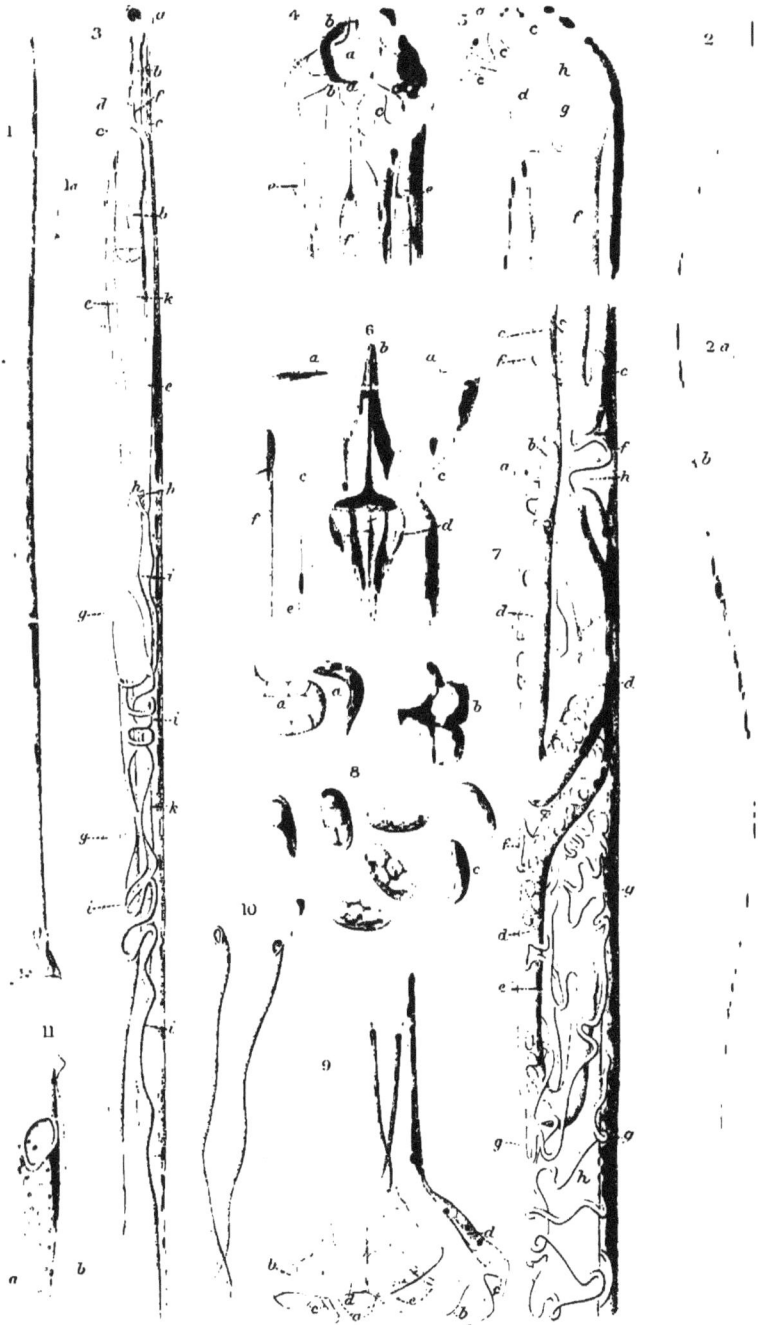

DOCHMIUS CERNUUS,
(The Bent-head Round Worm.)

Sclerostoma hypostomum, Dujardin.

Plate XXIII.

Description.—Male, 16^{mm}; female, 24^{mm}. Body white, cylindrical, and stout. Head, globular, a little ventrally curved and truncate at the mouth; head papillæ six; neck papillæ not observed; ventral cleft about opposite the end of the second fifth of the œsophagus, counting from the head. Unicellular glands and lateral lines very conspicuous. Mouth-opening round, obliquely inclined; surrounded by a single circle of very minute saw-like teeth. The globular chitinous capsule is marked by numerous longitudinal elevations, which are stronger at its bottom. The posterior opening is circular, unarmed, but roughened. Around the mouth is a circular canal which empties into a dorsal canal. There is apparently no pharyngeal apparatus as in *Dochmius.* Œsophagus slightly swollen caudally. Two lateral ducts opening near the mouth and situated on either side of the head are apparent; they seem to end caudally in the lateral canals. Male about two-thirds the length of the female. Bursa shallow, set on obliquely, campanulate. Ventral costæ either separate or slightly so; lateral slightly separate if at all; dorsal pair irregular and widely separate; the dorso-lateral and dorsal forms a single stem with three lateral branches. Spicula 1.5^{mm} long, linear aculeate, cross-striated, and bordered by narrow margins which roll in towards each other. Two anal papillæ. Female stout and thick, and usually with a brown crust near the vulva. Vulva very near the tail. Tail full to near the end, but ending in an acute mucronate point. Uteri directed toward the head; one, the caudal, makes a loop at the tail. The ovaries show loops near the tail. Eggs elliptical, 0.1^{mm} long, 0.06^{mm} broad, laid in the gastrula stage. Embryo not observed.

Occurrence.—*Sclerostoma hypostomum* is found in the large intestine of ruminants. It is present associated with *Œsophagostoma Columbianum* in sheep in this country, but not abundantly. It is closely related to *Sclerostoma equinum* of horses, a species said to make tumors in the intestines. As *S. hypostomum* is rare, it has not yet been determined whether it causes intestinal tumors in sheep. Since *S. tetracanthum*, whose embryos make tumors in the cæcum of the horse, is more nearly related generically to *Œsophagostoma* than to *Sclerostoma*, I am inclined to believe, as I have proven in regard to *Œsophagostoma Columbianum* in sheep, that the species belonging to the genus *Œsophagostoma* are intestinal—tumor-making parasites rather than those of *Sclerostoma*.

Life history.—M. Baillet (*Nouveau Dict. de Med. T. VIII, art. Helminthes,* 1886) is authority for the following:

The sclerostome of ruminants is reproduced in the same way as that of solipeds. Its eggs, of which the vitellus is segmented in the uterus of the female, are laid in the large intestine and carried without by the fecal material. They are hatched after a few days, and the embryos are very similar in their general form to those of *Sclerostoma tetracanthum*. They are cylindroid, subobtuse at the anterior extremity, and provided with a narrow tail much shorter than those of the young *sclerostomes* of the horse. They move, besides, in the same manner as the latter. They can live a long time in the fecal material of ruminants when they are not dried, and in this condition they grow. Some young *Sclerostomes* which, after hatching, were 0.35^{mm} long to 0.50^{mm} have been found in the droppings of sheep. After having been kept damp two and a half months these were from 0.66 to 0.78^{mm} long. Their skin, which is folded on the surface of the body, appears to indicate that, like the *Sclerostomes* of the horse, they are prepared to undergo a molt. The young *Sclerostoma hypostomum* can live a long time in water after having reached a definite size in the fecal material of ruminants.

We have not yet observed cysts within the mucous membrane of the large intestine of sheep.[*] But as the *Sclerostoma hypostomum* are in far less numbers of ruminants than *S. equinum* or *S. tetracanthum* in the horse, we can not yet draw any conclusion from the negative result of our researches. We will not say then at present whether the *Sclerostomes* of ruminants pass the second phase of their existence in cysts, or whether they are developed in the intestinal canal itself in the midst of alimentary material which is found contained there. The eggs of the *Sclerostomes* of ruminants taken directly from the uterus of females and preserved in water at a temperature of from 12 to 20° C. (about 55 to 70° Fah.) hatch at the end of four or five days after having undergone all the series of successive modifications which are alike observed in the eggs of *Sclerostomes* of solipeds.

The only possibility of error in Baillet's experiments is the indentification of the species with which he was dealing, for he writes:

It (*Sclerostoma hypostomum*) is above all very frequent in sheep. There are often found individuals, probably younger, whose mouth, entirely terminal, is less widely open, and provided with a single rank of teeth still less numerous. These worms also lack a pharyngeal capsule, and often carry a membranous swelling on the sides of the head.

The above describes *Œsophagostoma venulosum* quite well, and when the learned helminthologist did not experiment with eggs taken directly from the adult worm he may have had to do with the eggs of either. This would account for the similarities found between some of the young embryos experimented with and those of *Sclerostoma tetracanthum*. (The author is of the opinion that the last named species should be classed with the *Œsophagostoma* instead of the *Sclerostoma*.)

The life history of this species is, that the eggs are scattered by the sheep, that they then develop somewhat, and without the need of any secondary host are capable of further development in sheep when taken by them along with the food or drink, and that in the large intestine of the latter they may or they may not make tumors during their embryonic stages.

The prevention and treatment is the same as for other intestinal worms.

---

[*] This indicates that *Œsophagostoma Columbianum* has not been observed in France.— C. C.

SCLEROSTOMA HYPOSTOMUM.

PLATE XXIII.

Fig. 1. Male, ×8. Fig. 1a, natural size.

Fig. 2. Female, ×8. Fig. 2a, natural size.

Fig. 3. Head, ventral view: a, the capsule; b, the dorsal canal; c, lateral ducts; d, œsophagus.

Fig. 4. Head, lateral view: a, c, d, as in Fig. 3; e, the three head-papillæ; f, the common duct of the joined unicellular gland ducts is placed near the ventral cleft; g, the end of the œsophagus, showing three dependent lips; h, the cut-end of the skin. The drawing shows the longitudinal muscular band, separated into two groups by the lateral canals, i, i, and the three separations of the muscular bundles between these canals which appear on the surface as lines: k, k, k, cut end of the intestine.

Fig. 5. Head, dorsal view: a, b, c, and d as in Fig. 3.

Fig. 6. Mouth end of lateral half of chitinous capsule: a, a, a, the ventral, lateral, and dorsal head papillæ; b, the dorsal canal which continues around the head in c, c; d, the mouth, around which is a circle of small thorn-like teeth, about forty in all.

Fig. 6a. Teeth enlarged.

Fig. 7. Caudal end of capsule: a, the cut wall; b, the dorsal canal; c, the triangular opening of œsophagus.

Fig. 8. Section of œsophagus: a, a, a, the chitinous support; b, b, b, the walls, c, the orifice.

Fig. 9. Male bursa with spicula.

Fig. 9a. Portion of spiculæ: a, the chitinous cylinders; b, the membranous margins.

Fig. 9b. Diagram of anal papillæ.

Fig. 10. Bursa spread out: a, ventral costæ; b, lateral; c, dorsal; d, the torn edge.

Fig. 11. Caudal end of female: a, anus; b, vulva; c, c, uteri; d, d, loops of the ovaries; e, e, the intestine; f, a dark brown patch usually found on the vulva; g, an egg.

Fig. 12. Cross-section of female: a, intestine; b, ovaries; c, lateral ducts; d, muscular bundles (Leuckart).

Figs. 3, 4, 5, and 6 are somewhat distorted by flattening.

PLATE XXIII

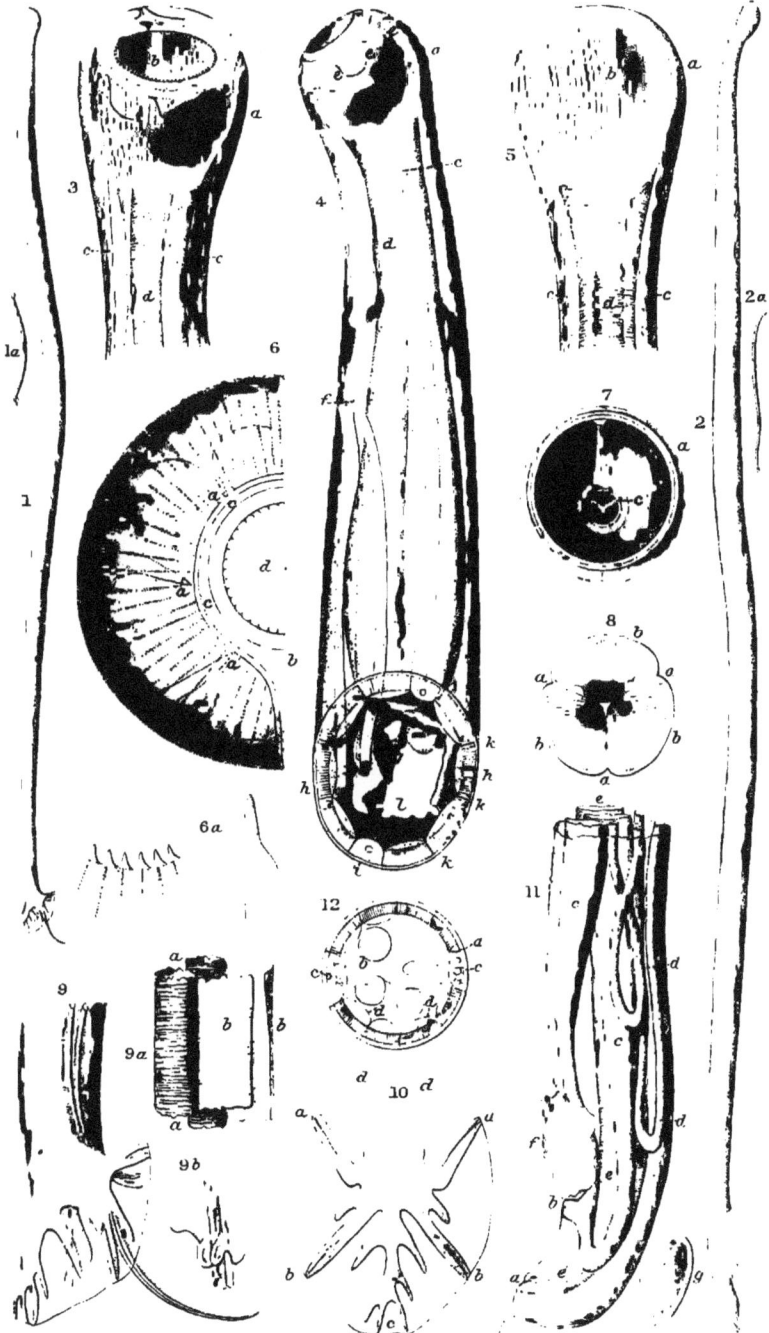

Haines, del.

SCLEROSTOMA HYPOSTOMUM

A. Hoen & Co. Lith Baltimore

THE NODULAR DISEASE OF THE INTESTINES.

ŒSOPHAGOSTOMA COLUMBIANUM, Curtice.

Plates XXIV, XXV, XXVI, XXVII.

In the Eastern States there exists a hitherto undescribed disease, which is characterized by tumors present in the upper part of the large intestine. The disease causes heavy losses, for it seriously affects the health of the sheep, and renders the intestine valueless for making sausage casings. Though the latter result would seem trivial at first sight, it is by no means unimportant, for sausage-makers are compelled to import the greater part of covering material used in their business. The disturbances of health produced are very serious, for there are places in the South where sheep can not be kept with profit, apparently on account of this parasite alone. Dr. D. E. Salmon, Chief of the Bureau of Animal Industry, who at one time lived in the South, performed many *post-mortem* examinations on diseased sheep, and found nothing but these intestinal tumors to account for the severe symptoms of disease which they exhibited, and has verbally stated that he believes this malady is the chief obstacle to successful sheep husbandry in some portions of the Southern States.

Investigation.—The cause of the disease remained until the winter of 1888-'89 in obscurity, but owing to a favorable combination of material and methods of investigation it was then ascertained. Some of the larger soft tumors, which are characteristic of this disease, were dissected from the intestine, and after being slit open their greenish, cheesy contents escaped in a watch-glass of water. By carefully teasing the apparently newer portion of these masses a little worm, the cause of the trouble, was found. Previous to that time Dr. Theobald Smith, of this Bureau, had in the winter of 1886-'87 made and examined microscopic sections of these tumors. One of a series of sections (see Plate XXV, Fig. 8) showed what was apparently the fragment of a worm. Numerous other sections made at the same time showed no signs of this parasite, and the investigations were temporarily abandoned.

Although tuberculosis is an uncommon disease in sheep, and although the tumors found in this disease differ in many essential points in both their history of formation and in their histological detail from those caused by tubercle bacilli, there was a superficial resemblance, on account of which Dr. Smith made many tests of the caseous matter for the bacilli both by the microscopical methods and by the inoculation of small animals. In the light of subsequent investigations it is needless to say that these tests had only negative results. In justice to Dr. Smith it should be stated that he undertook the experiments more for negative evidence than from any expectation of finding bacilli.

The cause of the disease is a nematode or round worm, which, though remarkably similar to some other worms of its group, is nevertheless a distinct species from any hitherto described. The failure to find this worm in our earliest investigations may be ascribed to two important factors. The first is that of the season of observation. We had probably overlooked those tumors which were best adapted to show the worms, but mainly on account of the season in which the examination was made. The method of examination was probably the greatest factor in the investigation, for as soon as what may be termed macroscopic superseded microscopic methods the discovery was easy. The method was not entirely macroscopic, for simple lenses of low power were used.

The worms found in the largest tumors were never over 3 or 4^{mm} in length, and presented only an embryonic development, the vegetative organs alone being present. Figs. 1, 2, and 4, in Plate XXV, were made from the largest specimens. As the adult worms of this class may differ materially from the embryos in the character of the mouth parts and in their appearance after the reproductive organs have developed, the difficulty was encountered of connecting the embryonic cystic form with an adult form existing in sheep or elsewhere.

Determination of adult.—The most conclusive method of determining the adult would be to directly develop an embryo into an adult, but this proceeding was not attempted on account of its difficulties. A less satisfactory method remained of finding some adult worm whose embryo was unknown, and which not only corresponded in structure with the embryo, but whose life history was such that it was possible for it to be the parent of the embryo. It will not be profitable to detail theories advanced to aid in this investigation; suffice it to say that while studying one day a group of worms which came from the large intestines of sheep, I found among them an undescribed species.

This species was immediately referred to the *Sclerostominœ* (Raillet, *Elements of Zoologie*, p. 330), and later to Molin's genus *Œsophagostoma*. There may be an impropriety in accepting this name over others proposed at the same time for species of this genus, but as the same author proposed them, and as the genus is in general use, it is accepted in this classification. For reasons assigned hereafter in a description of the species I have called it a new one—*Œsophagostoma Columbianum*. The specific name is from the fact that the worm was first found in the District of Columbia.

The adult worm is found in the large intestine of sheep in considerable numbers, and in the same animal may also be found the tumors.

The parasite is usually found below the narrowing of the large intestine, where the latter changes from a sack-like receptacle to a large tube, and below the mass of the tumors. The tumors may, however, extend the entire length of the intestine. It is evident that the distribution of the eggs of this adult would be favorable for sheep becoming again infected by them.

The oral armature of the embryo from the tumors and the adult *Œsophagostoma Columbianum* differ, that of the embryo appearing to be like a chitinous cup comparable to that found in the young of the *Sclerostoma* of sheep and the *Dochmius* of dogs. The young of these species, however, differ as much from their adult form as the embryo in question does from its supposed adult form.

The anatomical point that the writer considers of the most moment is that the embryos of the tumors possess a lapel-like fold on the ventral side of the worm just anterior to the ventral cleft, the opening of the unicellular gland duct, and that the adult *O. Columbianum* possesses a similar fold identically situated. (See Plate XIV, Figs. 1, 2, and 3, and Plate XXV, Figs. 2 and 3.) The absence of this feature in the embryos of *Sclerostoma* and *Dochmius* that infest sheep at once precludes the possibility that we have to deal with these species. That embryos of this character and corresponding adults are found in the same sheep makes the diagnosis more assured. Should we look for a host in some other animal we should expect to find one that had all the opprtunities of spreading the eggs that infect the sheep, but amongst our farm animals we find no corresponding parasite, and other wild animals which might be suspected as hosts are at the present time very rare.

ŒSOPHAGOSTOMA COLUMBIANUM, n. sp.

Description.—Male, 12 to 15ᵐᵐ; female, 14 to 18ᵐᵐ. Worm similar in appearance to *Dochmius cernuus*, but having its head bent into the form of a hook. Head terminal, very short, somewhat thicker than the neck, and separated from it by a constriction. Papillæ six, of which two are dorsal, two ventral, and two lateral. The latter are more obtuse and are the openings of the lateral canals of the body. Mouth terminal, supplied with a chitinous armature, consisting of an annular ring, which supports two systems of twenty-four teeth each; the outer circles are very long and curved, so that near their base they form a truncated cone and at their free extremities an inverted truncated cone. Within the outer is situated the inner and shorter row of bidentate teeth, so arranged that each tooth is opposite one in the outer row. Beneath the armature is the tri-radiate opening of the œsophagus. Neck not inflated, but provided with a lapel or fold of skin on the ventral side just in front of the ventral cleft; the fold continues slightly on to the dorsal side. Two lateral, narrow, membranous wings begin at this neck fold and continue for one-fourth the length of the worm. The two lateral opposite papillæ are in these wings, at the level of the first third of the œsophagus. The pair of unicellular neck glands unite into a common duct and empty at the ventral cleft, situated beneath the nuchal fold. Male about three-fourths the length of the female. Bursa saucer-shaped; can be spread symmetrically without tearing. Costæ or ribs symmetrical; the ventral slightly separated; the ventro-lateral forms with the lateral a group; the lateral also slightly separated; the dorso-lateral forms with its fellow and the dorsal pair a group; the dorsal rib is composed of two, which are widely separated only towards their free ends. Spicula two; awl-shaped, bordered by a very narrow membrane; the chitinous cylinder is apparently fenestrated. At either side of the genital orifice are two knobbed papillæ. Female relatively stouter; vulva just in front of anus, which is midway between it and the acutely pointed tail, usually covered by a hard brown patch; reproductive organs in two symmetrical sets anteriorly directed, except a small portion of one, which is at first posteriorly directed to form a loop in front of the anus. Uteri two, in the caudal end of the body. These may be traced up to the oviduct and ovaries, which continue to the cephalic end of abdomen, where they are reflected to form a loop and thence continue to the caudal end to form still another loop. The ovary of the caudal uterus forms still another loop nearly opposite the uteri.

Eggs laid with gastrula inclosed; length, 0.09^{mm}; width, 0.05^{mm}. Embryos from 0.23^{mm} upwards. The largest found in tumors were 4^{mm} long; the smallest male found in intestine was 7^{mm} long. The smallest embryos were without digestive apparatus. The largest possessed an intestine, unicellular glands, and a well-formed chitinous spherical cup in the head; also six cephalic papillæ, and at the neck two papillæ and a well-defined fold.

Occurrence.—The adults live in the large intestine of sheep below the cæcum; the embryos live in the intestinal walls in tumors, which, though more abundant in the cæcum, may be scattered from the duodenum to the anus. The species is distributed in the United States east of the Mississippi River as far north as Maryland, perhaps farther north. On comparing the species with others of the genus it was found that it corresponded more nearly with *Œsophagostoma venulosum*, a parasite of goats, than any other described in modern works, and that it corresponded still more closely with *O. acutum*, Molin. (*Il sottordine degli acrofalli. Memorie dell * * * Istituto Veneto*, 1860, *Vol. IX, p.* 449.) The latter is apparently a distinct species, although in quoting other authorities he has given an incorrect synonymy. *O. Columbianum* differs markedly from *O. venulosum* in not having an inflated neck, and from the latter and *O. acutum* by the possession of a lapel of skin upon the ventral side of the neck, just in front of the opening of the unicellular neck glands. Molin examined two males and three females of *O. acutum* from *Antilope Rupicapra;* one male from *Capra Hircus*, var. *Mambricus*, in association with forty-five *Sclerostoma hypostomum;* and two males and one female from *Capra Ammon*. The finding of *Sclerostoma hypostomum* in association with the *Œsophagostoma acutum* is the fact which leads me to infer that the latter occurred in the large intestine of the above mammals and not in the stomach, as he cites in his synonymy of the species.

Life history.—*Œsophagostoma Columbianum* seems to have become a parasite of sheep since their introduction into this country. If present in the Old World at all it is sparingly so, and seems to have escaped detection. So little is known about its distribution, that it is impossible at present to accurately define its limits. From its great abundance in the Southeastern States one might infer that it had originated as a sheep parasite in that region, and probably from some animal of allied organization and habits. The writer is at present inclined to believe that the deer may have been its host at the time European sheep were first introduced. There are few facts to sustain this hypothesis. Though it may yet be too early to form a positive conclusion, further investigation may determine the section of the world to which this parasite originally belonged and then the former host may be indicated with tolerable accuracy.

The life history of this parasite seems to be completely known to us during its development from the immature form found in the intestinal tumors to the adult stage; but there is a period from the moment when the eggs escape from the intestine with the excrement to the time when it is found again in the intestinal tumors that must remain in obscurity. After the embryo has returned into the alimentary canal it makes its way through the mucous coat of the intestine and becomes encysted there. The writer has been unable to learn how it passes through the mucous coat, as even on the youngest specimens no sign of armature is found. The very young forms found in the cysts show little differentiation beyond what they could have attained in the eggshell. They are soon surrounded by a cyst which seems to belong to them and to have been created by them; but whether this cyst is the remains of a molt or not can not be asserted. Later in their history they become surrounded by the products of the inflammation they excite in the surrounding tissues, and eventually break from the cyst and live in the cheesy mass of the tumor. In this stage of their growth the worms exhibit the intestine and oral cup and indistinct unicellular glands. They then molt, and show all these features in more distinct outline. Having attained a length of from 3 to 4^{mm}, or less than one-sixth of an inch, they break from the tumors to begin their life in the intestine. In the latter they continue their growth and becoming sexually perfect and produce eggs which eventually go through the same cycle.

In developing, this worm molts at least three times—once in passing out of the stage in which it has no mouth or intestines, once during the development of these parts as we find them in the embryo, and once while the worm changes from the embryonic form to the adult form.

Disease.—The harm that these parasites do the sheep is directly dependent on their numbers and life history. Yearlings may show considerable infection, but it is usually in older sheep that the most abundant infection occurs. The disease is a seasonal one, in that it can be found in best development in the winter. The lambs begin to be infected in the summer and fall, and from that time the tumors formed increase in size until early in the spring of the next year, when they gradually grow smaller but probably do not entirely disappear.

Pathology.—A study of the fresh tumors by compressing the smaller ones between two glasses and by dissecting larger specimens gives the following results: The small tumors, which are scarcely the size of a pinhead, are found in the submucous tissue. They appear like a sac filled with fluid and having a little globe floating within. By using a higher magnifying power the little globe is seen to be a cyst with a worm inclosed (see Plate XXV, Fig. 6). By careful dissection the cyst may be separated (Fig. 5), and finally the worm itself may be separated (Fig. 3). At this stage the worm shows little differentiation of parts. In examining another and older cyst (Plate XXV, Fig. 7) the same appearance may be observed. There is also a little greenish cheesy substance present. A third stage (Plate XXVI, Fig 3) shows the latter still further increased, and in the figure referred to the arrow shows that the encysted worm has been pressed out of this mass, leaving a cavity behind.

When the little tumors become larger than a pinhead and entirely filled with the cheesy matter their structure does not materially change, but is more difficult to make out. It is at this stage that the worm escapes from the cyst and begins to wander within the capsule which its presence in the tissue has caused. On dissecting the large fresh tumors the worm is found in the mass of cheesy material, which is now quite abundant, filling the cavity and producing tumors as large as the end of one's finger. This cheesy material is usually hard, dry, and brown at one end, and soft, yellowish-green at the other. It is in this end that the worm is to be found. If some of the harder tumors are examined, it may be found that there is no greenish material in them and no worms. These hard tumors may be of all sizes and are found at all times. From these the worm has either escaped or in the case of the smaller tumors the worms have died. The form of these tumors is usually spherical, but the cheesy material may appear as a long mass, or it may apparently fill what seems to have been a worm track. The last appearance occurs most often in the small intestines. From the mucous side these tumors present little if any color. The older ones may present a greenish-yellow appearance, especially if the mucous membrane over them be thin. In well-advanced cases, when the tumors are numerous and large, many are found in which the mucous membrane over them is ruptured and the cheesy mass protrudes into the intestinal cavity. In these no worms have been found, and hence the conclusion has been reached that the worms have escaped.

In lambs the little dots indicating the presence of the young tumors are very scattering (see Plate XXVI, Fig. 1), but in older sheep they may be very numerous (see Plate XXVII, Fig. 1). Between the stage in which a few are scattered here and there over the cæcum and that in which the cæcum has become a stiff tube with walls from one-fourth to one-half inch in thickness (so thick and close have the tumors become) there are all varieties. The tumors may extend from the cæcum to the anus. They may also be abundant along the small intestine. The cheesy material which the worms produce has been found in the lymphatics, on the omentum, and in the liver, but in these places it never seems to be sufficiently abundant to show that the parasite lived long.

Microscopic serial sections made from alcoholic specimens show best the changes which have taken place in the surrounding tissue. On Plate XXVI, Fig. 2, there is figured an enlarged view of one of the worms still encysted and surrounded by the tumor of inflammation. From this section it is seen that the irritation set up in the adjoining connective tissue causes the cells to proliferate and crowd closer and closer together. It is also seen that there is a special cyst for the worm and a thickened adjoining portion of the tumor which is like a surrounding membrane. As these two membranes, the so-called cyst and the outer one, stain so nearly alike, it has occurred to me that they represent successive efforts of the adjoining tissue to protect itself against the parasite; but in view of the fact that the inner one is so easily enucleated, it is for the present considered as belonging to the worm. At the foot of Plate XXV, Fig. 8, there is given an illustration of a small tumor in which the worm has escaped from the cyst and in which the surrounding matter has become cheesy, some of it even hard. Around the entire mass the tissue is slightly thickened and forms a capsule.

From these two microscopic sections we can learn how these tumors grow. The worm penetrates to the submucous tissue and irritates it. The adjoining cells rapidly increase in number and crowd upon each other. So closely do they crowd and so numerous do they become, that the outer layers cut off the circulation from the inner cells and they die. Their degeneration gives rise to the cheesy mass. Now, if the worm remained in the center of th e mass the new growth would eventually cease, but the worm makes its way to the outside and at that point keeps up this irritation and new growth. This is shown by and accounts for the dried older parts of the larger tumors and the fresh yellowish-green adjoining parts. As soon as the worm escapes the irritation ceases, the tumor shrinks, and absorption of the mass begins. The irritation produced by the worm provides it with food and favorable surroundings for development. Often the worm dies from weakness or other cause, and leaves behind those little hard incompletely-grown tumors which have been mentioned.

Since writing the above life history two *post-mortem* examinations have been performed, which lead me to modify my views regarding the life history of certain other tumor-making parasites. On August 10, 1889, two lambs, one five months old, the other eleven weeks old, died. The older lamb was bred at the experimental farm; the younger was bought with its dam when but a day or two old. These lambs had no water save what was pumped from a well. Among other parasites, each species being found in its proper portion of the intestine, there were numbers of individuals of *Œsophagostoma* scattered through the length of the large intestine. These individuals were of all sizes, from the smallest stage (7^{mm} long) to those nearly adult. In the younger lamb there were but few very minute tumors in the coats of the large intestine. In the older the tumors were somewhat larger, but none were much larger than a millet seed. The abundance and size of the free parasites indicated that most if not all of them had developed in the interior of the tube. The intestinal tumors indicated that there were others developing in them.

From the above it seems probable that this species develops normally in the intestine, that some of the young embryos penetrate into the walls of the intestines, and at times even to the mesenteric lymphatic glands and elsewhere; that those which penetrate into the intestinal wall either develop very slowly or eventually die, depending on the tissue penetrated and the favorable conditions the latter offer for the nourishment of the parasite; that those which develop slowly may in time escape from the tumors and complete their development in the intestinal canal; that this phase of its life history may be one that is favorable for the preservation of the species by preserving a few individuals in tumors throughout the winter which escape in early spring, become adult, and lay their eggs, which are scattered on pastures favorable to the preservation of the embryos; that the majority of these embryos, penetrating the intestinal walls, wander into such environments that they are eventually destroyed; that this act of migration is voluntary and only of benefit to the parasite when the latter becomes lodged in the proper place, and that the same power which enables it to arrive at these places also enables it to penetrate farther than is of use to it; that the slow development of the worm in the tumors as compared with the development in the intestine shows it to be a retarded development, which may be hastened as soon as the parasite again reaches the proper surroundings.

In short, the fact of the intestinal life of these parasites is demonstrated, as well as the fact that some may wander into the intestinal walls and undergo a retarded development before re-entering the intestine.

The diagnosis of this parasitic disease can only be made from a *post-mortem* examination. In the living sheep there may be signs of general debility—bloodless lips and eyes, thin sides and flanks, dry wool, etc. It may be that nothing else will be noticed, but that the flock is not in quite good condition; or in severe cases the diarrhea and emaciation may be excessive. Dr. Salmon believes the disease may bring death to its victims in the severest cases. My own observations have been confined to the abattoirs, where salable animals only are brought. As the adult worms are comparatively few as compared with the tumors, it is probable that the adults of this species cause but little trouble; but the embryos, on the contrary, cause a great deal. The disease is an insidious one, for not only is the rate of infection gradual, increasing slowly in amount from week to week, but the rate of development of the tumors is very slow, apparently requiring months. It is only when the disease is well advanced that its cumulative effects can be noticed.

The disturbance of digestion caused by this parasite is mainly due to the derangements of the functions of the *cæcum*. This derangement is not serious until the resulting tumors become exceedingly numerous, well advanced in growth, and press upon the more essential mucous membrane, disturbing its functions.

The most seriously affected sheep found in the abattoirs are noticeably poorer, and one would be tempted to believe, were he to judge from the "knotty" viscera, as the butchers call them, that such animals should have died from the disease long before. These sheep usually have diarrhea, a disease which weakens the affected animals. Flockmasters who mistrust that their sheep are not doing well, and who know of no cause for it, should sacrifice one or two of the poorest to make a diagnosis. The meat of such sheep, though not quite as fat as other mutton, is suitable for food, and could not be distinguished in the market from other mutton.

Prevention.—For the tumors caused by *Œsophagostoma Columbianum* there is no remedy except the removal and extermination of the adult worms. These adults are usually buried deeply in the mucous secretions and attached to the membrane of the large intestine. They may be found in considerable numbers in older sheep. The medicinal remedy must therefore be one which will remove them from these places. It is probable that such a one can eventually be found, but at present none can be recommended. It is probable that some one of the surer remedies advised for other intestinal parasites will do for these. In case medicinal remedies are tried each animal must be dosed. The killing of the adults will of course lessen the number of eggs with which the sheep become infested. As the eggs of this parasite pass to the ground the sheep may get them either while pasturing or drinking. The same care in changing pastures, in providing good drinking water and a plentiful supply of salt, should be observed as for other parasites. Judicious fall and winter marketing of infected sheep will also tend to lessen the chances of infection. If pastures are known to be permanently infected, then they should be turned over to other stock for a year or two before being again grazed on by sheep. When it is practical on the smaller farms the sheep lots should be plowed and either planted or left fallow. The object of change of pasture and of plowing is nearly the same; in the one case, to wait until the parasites have died out; in the other, to bury them beneath several inches of soil, from which the sheep-owner may rest assured they will not emerge.

ŒSOPHAGOSTOMA COLUMBIANUM, Curtice.

PLATE XXIV.

Fig. 1. Adult male, ×9, Fig. 1a, natural size.

Fig. 2. Adult female, ×9, Fig. 2a, natural size.

Fig. 3. Cephalic end of adult, ventral view: a, head; b, œsophagus; c, lateral canals opening on head; d, unicellular glands and ducts uniting and emptying at e, the ventral orifice; f, a fold of the skin which forms a half ring on the ventral side of the neck; the orifice e opens nucer it; g, g, two pointed neck papillæ; h, h, narrow membranous wings; i, intestine.

Fig. 4. Caudal end of adult female lateral view, ×45 (about): a, anus; b, vulva; c, vagina; d, d, uteri; e, e, oviducts; f, f, f, loops of the ovaries; i, intestine.

Fig. 5. Head: a, mouth; b, b, papillæ of the two lateral ducts; c, c, two of the four acute, pointed papillæ opening on the head; d, d, cross-section of cavity of the head; e, e, cross-section of a circular canal, which runs around the base of the head; f, chitinous armature of the head; g, œsophagus; h, its three chitinous supports.

Fig. 6. Chitinous armature, top view: a, chitinous ring; b, outer teeth; c, inner teeth; d, triradiate pharyngeal orifice; e, e, e, the ends of the œsophageal rods. Four of the outer teeth have been removed to better show the inner row. A side view of this apparatus is presented in Fig. 5.

Fig. 6a. Top view of teeth.

Fig. 6b. Side view of teeth and basal support.

Fig. 7. The bursa spread out: a, ventral rib; b, ventro lateral; c, lateral; d, dorsolateral; e, dorsal; f, 'the cloacal orifice with two papillæ behind.

Fig. 8. Spicula of male, composed of a chitinous tube surrounded by a membrane with a narrow wing.

Fig. 8a. Portion of spiculum.

Fig. 9. Eggs: a, a, as they appear when deposited; b, eggs in uterus.

Fig. 10. Bursa of male, side view: a, a, a, ribs; b, spicula; c, c, anal papillæ.

Fig. 11. Head, top view: a, mouth; b, b, lateral papillæ; c, one of the four acute papillæ.

PLATE XXIV

OESOPHAGOSTOMA COLUMBIANUM
(The Tumor-Making Round Worm of Sheep.)

ŒSOPHAGOSTOMA COLUMBIANUM, Curtice.

PLATE XXV.

Fig. 1. Worm in third stage, ×60: a, head, with chitinous armature; b, œsophagus; c, intestine; d, unicellular glands; e, anus; f, line denoting natural length.

Fig. 2. Cephalic end, lateral view, ×150: a, chitinous cup; b, one of the six cephalic papillæ, (these are slightly distorted); c, side view of the neck-fold, under which the gland-ducts d, d, open on the ventral line; e, œsophagus; f, intestine.

Fig. 3. Worm in first stage when 0.23ᵐᵐ. long. No internal organs were seen in this specimen.

Fig. 4. Cephalic end, ventral view: a, head; b, neck-fold, near, which are the neck papillæ g, g; c, œsophagus; d, d, unicellular glands, which open under the fold b; e, intestine; f, f, glands.

Fig. 5. Worm inclosed in its cyst.

Fig. 6. Tumor from which the cyst in Fig. 5 was taken: a, surrounding tissue dissected from cæcum; b, fluid-filled space; c, capsule with inclosed worm.

Fig. 7. Older tumor. This differs from tumor of Fig. 6, in having a movable cheesy mass. These tumors are distorted by compression, which makes them too flat. Figs. 6 and 7, and Fig. 3, Plate XXVI, present three stages in the disease, as seen in fresh tissues under low magnifying powers.

Fig. 8. Section of an advanced tumor: a, mucous membrane; b, submucous; c, inner muscular layer; d, outer: e, serous membrane; f, the cheesy mass of the tumor in which is a small section of the worm. This presents a more advanced stage of the disease than Fig. 2, Plate XXVI.

PLATE XXV

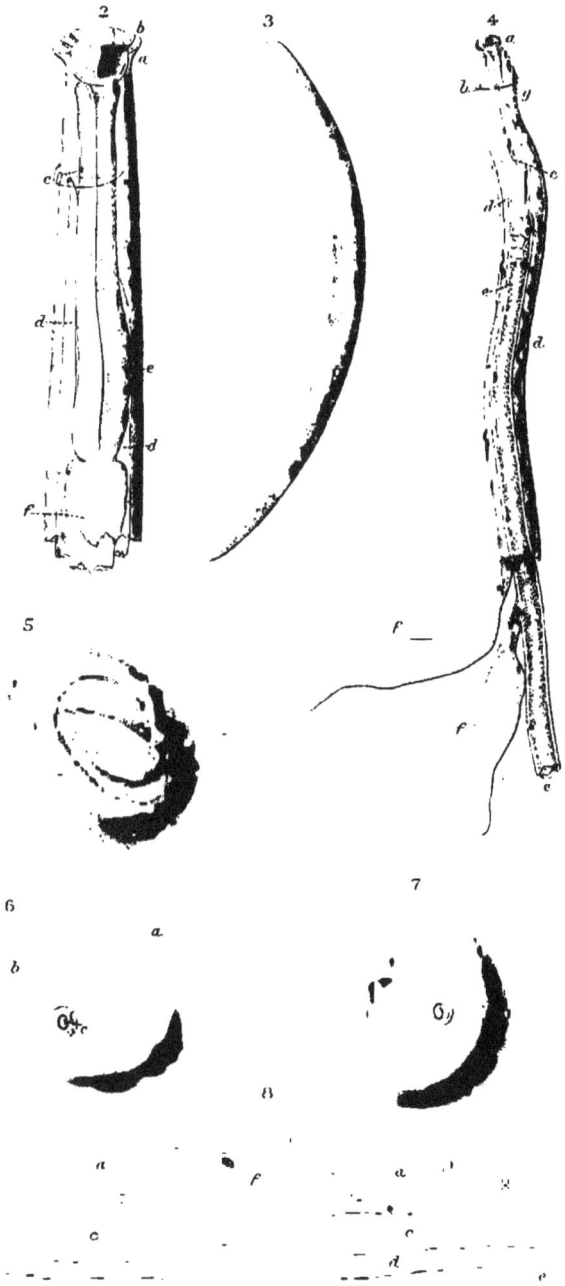

ŒSOPHAGOSTOMA COLUMBIANUM
(Young Stages.)

ŒSOPHAGOSTOMA COLUMBIANUM, Curtice.

PLATE XXVI.

Fig. 1. A piece of mucous membrane taken from the cæcum, surface view, natural size. The patch of dots scattered uniformly over the surface represents intestinal glands; the irregularly scattered larger dots and elevations are the young worm tumors in their first stages.

Fig. 2. A section through a worm tumor in its younger stage: a, a, a, mucous membrane; b, submucous connective tissue, in which are c arteries and d veins; e, the tumor, which is made of connective tissue cells and their nuclei, packed closely together; near its center is the worm cavity f, with a piece of the worm, which is surrounded by a section of a special capsule; outside of this is a thick membrane, formed from the surrounding material.

Fig. 3. Small tumor dissected from the cæcum: a, the outside capsule filled with fluid, in which is b, a hard, cheesy mass; c, the worm in its capsule, which has been pressed out of the cavity in the mass b. This presents a more advanced stage of the disease than fig. 7, Plate XXV.

Fig. 3a. The ruptured capsule.

Fig. 3b. The worm near the end of its second stage about to moult.

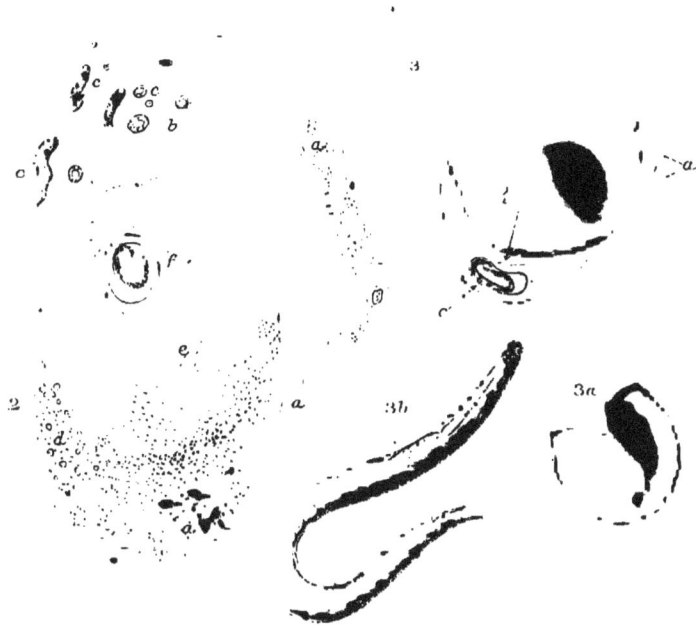

PLATE XXVI

ŒSOPHAGOSTOMA COLUMBIANUM,
(Small Tumors of Cæcum.)

ŒSOPHAGOSTOMA COLUMBIANUM, Curtice.

PLATE XXVII.

Fig. 1. Piece of cæcum exhibiting tumors caused by the embryos of *Œsophagostoma Columbianum*, natural size. The various stages of growth are represented by the different sized tumors. The smallest are better shown in Plate XXVI, Fig. 1.

Fig. 2. Cross-section of Fig. 1, at *a a*; *b*, mucous membrane; *c*, submucous; *d*, muscular and serous layers; *e e e*, section through the cheesy masses.

2

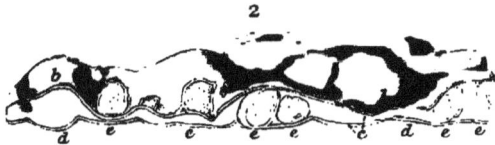

ŒSOPHAGOSTOMA COLUMBIANUM,
(Large Tumors of Cæcum.)

THE CÆCUM WORM.

Trichocephalus affinis, Rud.

Plate XXVIII.

Description.—Male and female about equal, 40 to 70ᵐᵐ long. Body whip-like, possessing a short, stout caudal end, 12 to 18ᵐᵐ long, and a very thin hair-like cephalic end of twice this length. The latter contains the œsophagus and intestine; the former the reproductive organs and intestine.

The head is very small and thin, without noticeable papillæ or chitinous armature. It is said to sometimes have two vesicular, transparent, wing-like inflations. Skin of the neck transversely striate, and when highly magnified shows a serration of the sides indicating cuticular layers which overlap each other like shingles on a roof. Œsophagus and cephalic portion of intestine very minute; its posterior end is large and dark, and empties at the caudal end of the body. On one side of the head there appears to be a canal filled with granules.

The male is to be distinguished by its tightly-curled caudal end. The testicle, beginning near the caudal end, continues anteriorly as a sinuous tube for about two-thirds the length of the thick portion of the worm; it then becomes plaited to the end of the thick part, where it turns and continues posteriorly as an enlarged seminal duct for about half the length of the thickened body, where it is constricted; the remainder continues to the cloaca as a slightly enlarged tube. The intromittent apparatus consists of two parts, an external membranous tube bristling with spines and an internal long, slim spiculum. It is always found exserted, and usually has one coil in it. The tube shows at its end that the external covering continues around the end into the tube to form a lining membrane, which may be retracted or protruded. There is considerable space between these membranes at the tip, and it assumes various forms, varying between a large sphere, as shown in the figures, and an elongate cylindrical body. The chitinous spiculum is terminated by an acute point. It is from 5 to 6ᵐᵐ long, with a width of 0.025ᵐᵐ. The tube is about three or four times as wide. The spiny points are turned away from the end.

The female has a thick body, only slightly curved. Tail, obtuse; ovary begins at the caudal end, continues as a plaited canal to the cephalic end of the thick part of body, then contracting returns to the caudal end where it enlarges, forms a fold, and becomes the uterus, which empties through the sinuous vagina and the vulva at the cephalic end, where the body begins to enlarge. Eggs characterized by having refrangent polar bodies at each end. They measure 0.077ᵐᵐ in length, including these bodies, or 0.056ᵐᵐ excluding them (Raillet). They are elliptical and dark brown.

Occurrence.—This species is found in the cæcum of sheep, goats, and cattle. When the fresh intestine is examined the worm may be found with its slim, hair-like head firmly sewed into the mucous membranes. The serrated structure of the skin not only facilitates the progress of the head through the mucosa, but prevents its being pulled backward. The thick large end, which is what one really sees at first, appears to float free in the intestinal contents.

The life history of this species has been determined by Leuckart, the distinguished helminthologist, who has added so much to this branch of biology. He succeeded in raising young embryos from the eggs to such a stage that there was no reasonable doubt that the next stage was passed in sheep. These he fed to a lamb, which he killed after sixteen days. In these he found numerous immature *trichocephali* about 1ᵐᵐ in length. He later verified this experiment by another, with like results. (*Die menschlichen Parasiten*, Band II, 494–499.)

These experiments show that the eggs of *Trichocephalus affinis*, which pass from sheep to the ground, may develop there to some degree, and then, after being consumed with food or drink by a second sheep, continue their development to their adult stages.

Disease and treatment.—Unless the parasite should be present in great abundance the species does not seem to be especially harmful. A few may be found in nearly all lambs and young sheep, especially in the all. The means of prevention is just the same as for other *round* worms. As they are attached so stoutly to the mucous membrane it is doubtful whether medicinal remedies would have the influence on them that they have on those worms situated in the small intestine.

TRICHOCEPHALUS AFFINIS, Rudolphi.

PLATE XXVIII.

Fig. 1. Piece of cæcum with *trichocephali* attached, natural size: *a, a,* females; *b, b,* males.

Fig. 2. Male, ×7: *a,* capillary cephalic end; *b,* coiled caudal end; *c,* protruded intromittent organ; *d,* the convoluted, and *e,* the straight portion of the seminal apparatus; *f,* seminal reservoir; *g,* intestine.

Fig. 3. Female, ×7: *a,* capillary cephalic end; *b,* vulva; *c,* vagina; *d,* uterus; *e,* oviduct; *f,* convoluted ovary; *g,* intestine.

Fig. 4. Caudal end of male enlarged: *a,* end of the body; *b,* spine-covered tube of intromittent organ; *c,* its inflated end; *d,* spiculum.

Fig. 5. Cross-section of end showing how the outside sheath becomes converted into the inside sheath of the tube: *a, a,* the sheath; *b,* the sac formed; *c,* the hollow spiculum.

Fig. 6. End of sheath, much enlarged, to show the relation between sheath and spiculum.

Fig. 7. The head.

Fig. 8. The vulva and vagina, with an egg in the passage.

Fig. 9. Eggs: *a,* eggs without shells; *b,* egg with shell and its characteristic polar bodies; *c,* intermediate between *a* and *b.*

Fig. 10. Enlarged portion of worm from near the head.

PLATE XXVIII

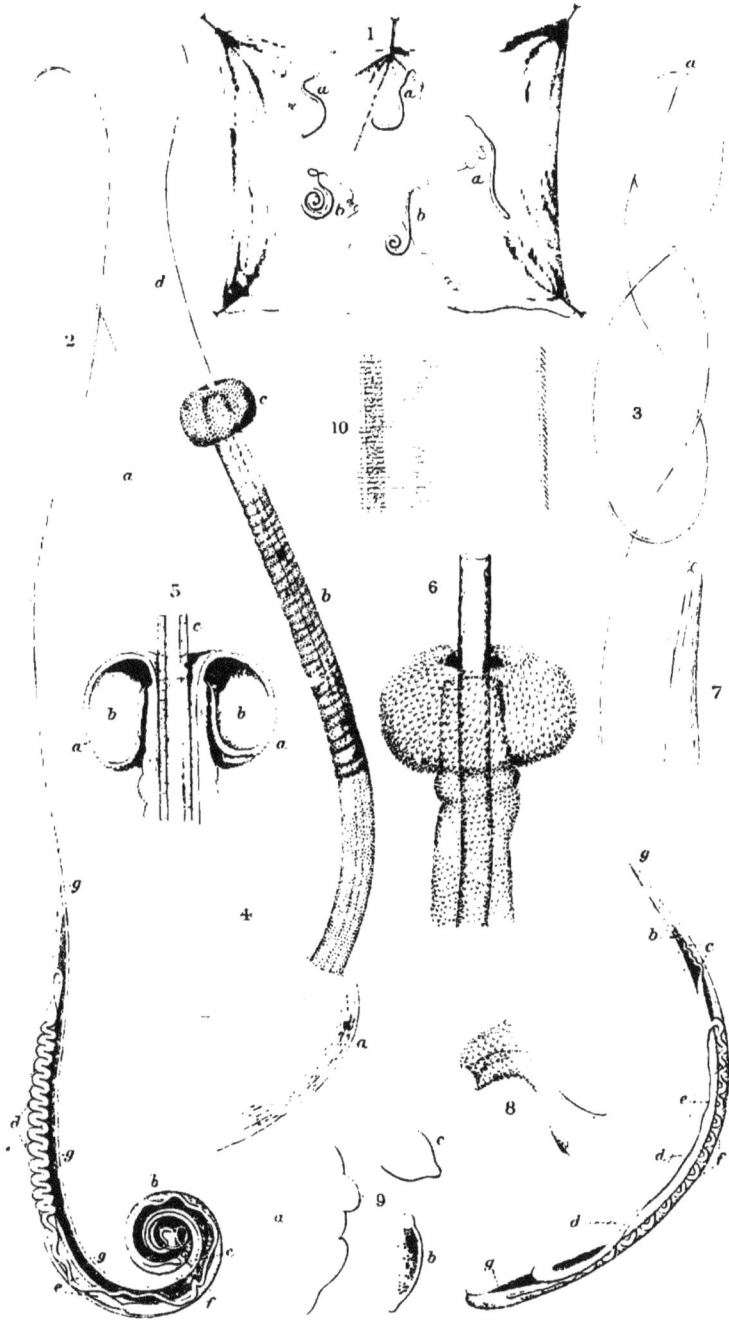

Gen. Marx, del.

TRICHOCEPHALUS AFFINIS,
(The Hair-headed Round Worm.)

A. Hoen & Co. Lith. Baltimore

LUNG-WORM DISEASES.—PAPER SKIN, HOOSE, HUSK.

VERMINOUS PNEUMONIA—VERMINOUS BRONCHITIS.

Plates XXIX to XXXVI.

The lung worms which cause disease in sheep in the United States belong to two, perhaps three, different species. They are *Strongylus ovis-pulmonalis*, Diesing, the hair lung worm ; and *Strongylus filaria*, Rud., the thread lung worm. The third species, which has been reported as infesting sheep in Europe, is the hog lung worm, *Strongylus paradoxus*, an abundant species occurring in the lungs of swine in this country, and while it has never been credited as having been found in our sheep it is to be looked for. As its size and the disease it causes is similar to that of *Strongylus filaria*, it will not be treated separately.

The diseases produced by these species of worms are caused by the mechanical injuries the worms inflict on the delicate membranes of the lungs and the clogging up of the air passages by them and the débris which they produce. The two forms of disease produced depends on the different size and habits of the two species. *Strongylus ovis-pulmonalis*, being very small, penetrates the air passages to their endings in the bronchioles and infundibuli, and causes disease in them primarily, while *Strongylus filaria*, which penetrates only into the bronchi, creates a disturbance there which produces a solidification of the lung secondarily. The general diagnosis of each disease is no easy matter. The disease produced by *Strongylus ovis-pulmonalis* is characterized by the spongy feeling of the lung and the presence of nodules from the size of a mustard seed to that of a pea scattered under the surface of the dorsum of the lung and at its posterior free edge. These may be connected by a grayish, fleshy, intermediate portion of the lung into patches of considerable size. The parasites can scarcely be seen by the unaided eye ; but if small pieces of the affected lung or a tubercle be placed in a shallow dish of water and teased out with needles under a tripod lens, they can be readily seen.

The disease produced by *Stronglyus filaria* and *S. paradoxus* is characterized by the posterior portion usually, or some entire section of the affected lung appearing as a solid, usually red, mass which has lost all of its contained air and is in a state of hepatization. A piece cut out generally sinks in water, while pieces from the former disease float. If in the latter disease the trachea is carefully slit open and the branches

185

traced down to the affected part the parasites will be found in thread-like bunches, completely filling the tubes.

The symptoms of lung-worm disease in sheep can not well be diagnosed in living animals unless the disease is far advanced, and then only in the severer cases. Sheep affected with either disease generally have pale, bloodless mucous membranes, harsh, dry hair, a dejected look, more or less difficulty in breathing, and often a deep cough. The bloodless condition of the sheep could arise from other parasitic troubles, but the disturbance of respiration should lead one to suspect lung parasites.

Consumption or tuberculosis is apparently a rare disease in sheep, and is not liable to be confused with this disease, which can always be diagnosed by finding the parasite. Lung-worm disease differs from acute bronchitis or pneumonia in being of slow development, and is less severe in its symptoms. Worm diseases consume weeks in development, while acute diseases are begun and finished in a few days.

THE HAIR LUNG WORM—VERMINOUS PNEUMONIA.

STRONGYLUS OVIS PULMONALIS, Diesing.

Plates XXIX, XXX, XXXI, XXXII, XXXIII.

Description.—Male, 16^{mm}; female, 25^{mm}; width, male, 0.5^{mm}; female, 0.17^{mm}. Capillary integument of worm very transparent, the cavity of the body appearing as a dark line. Head not winged; four papillæ; mouth naked. Male, bursa pointed, compressed, terminal; costæ (apparently) seven; one posterior; twice-notched; two pair lateral, one pair anterior; spicula symmetrical, spatulate, curved; 0.15^{mm} long, divided into two nearly equal parts; the anterior consisting of a cylindrical chitinous skeleton with a membraneous expansion, the posterior of a transversely ribbed skeleton, margined by a thin broad curved membrane, the two spicula forming a partially closed tube. Female oviparous, with two uteri and ovaries. The former empty into a vagina at 0.8^{mm} from vulva. Vulva 0.1^{mm} from anus. Anus 0.08^{mm} from tip of tail. Tail ends in a blunt point. Eggs in uteri 0.1^{mm} long, 0.04^{mm} wide. The eggs segment after being laid. Embryo provided with a very sharp-pointed tail.

Life history.—The young of the hair-lung worms escape from the lungs of infected sheep and become scattered over the pastures, yards, and other places frequented by these animals. They are then taken with the food or drink and in some way arrive in the lungs of the sheep. Arriving at the extreme ends of the bronchial tubes, they break down some of the tissues and become encysted. In the cyst they grow to adult size and take on sexual characteristics. Escaping from the cysts they make their way into the small air-tubes (bronchioles and bronchi), where the sexes mate and reproduce. The eggs are then laid in surrounding cavities and hatched into young worms, which make their way into the neighboring air-chambers (infundibula). Afterwards some of these worms may be coughed out of the lungs onto the pastures and infect other sheep.

In their life history there are but one or two points about which there can be any question. Many learned helminthologists believe that the young worm must escape from the sheep in order to spend a portion of its life on the ground or in some of the minute forms of animals before they are capable of further development in the sheep. Most authors are agreed that the worm passes into the lungs by the trachea either during feeding or rumination. The length of time which it takes the worm to complete its cycle of life is yet unknown. As the most pronounced cases among slaughtered animals are in the older sheep, it would seem as though this parasite was of very slow growth, requiring years instead of months for successive generations to produce a disease fatal to the infested sheep. It may be, however, that many lambs and young sheep are so seriously affected with the parasite that they either die or become so inferior in quality that they are never taken to the abattoir. In this case the cycle of life would prove to be rapid.

Disease.—The diagnosis of verminous pneumonia in living animals is a difficult matter. Not until the disease is so far advanced that its cure is hopeless are any well-pronounced symptoms developed. The worst affected sheep may have a deep cough, be out of condition, and be generally anæmic, as shown by the pallor of the visible mucous membrane and the dry, harsh coat. They are likely to lose flesh, but some, if not seriously affected, fatten tolerably well.

The *post mortem* diagnosis is as certain and definite as the diagnosis in life is unsatisfactory. So pronounced are the lesions caused by the worms in the lung tissue that any one having once seen a diseased lung would easily recognize it again. The little tubercles, filled with greenish material and surrounded by more or less of the thickened lung tissue which when cut exudes a frothy liquid, are diagnostic. The presence of the worm in these tubercles is decisive.

The *prognosis* of this disease can not be definitely given. From many examinations of affected lungs it seems to me that the disease is a progressive one, producing its worst effects as the sheep grow old. Where the sheep are marketed young the loss from this parasite is comparatively small; but where the disease is wide-spread and affects whole flocks, though but a few cents may be lost per head, the aggregate loss to the sheep industry must be considerable. To this must be added the loss from the disease in its more severe stages. When the disease is once in a flock and the farm or range is infected with it there will be a steady loss resulting until the disease is in some way exterminated.

Pathology.—The disease created is dependent upon the life history of the parasite as to character and upon the numbers of the invading hosts for its intensity. The changes produced in the lungs are but the aggregate of all the changes which result from the different invading individuals, and the history of the changes wrought by a single parasite illustrates the changes produced by all. The minute worm, when entering, penetrates the air passages to their extremities. In the ultimate alveoli it breaks down some portion of the membranous partition and becomes surrounded by the products of the inflammation which it excites and forms a very minute tubercle. When this tubercle has reached from one to two and one-half millimeters (one twenty-fifth to one-tenth of an inch) in size, it is composed of a distinct central part, filled with a soft, greenish, central portion, which is surrounded by a thicker membranous capsular portion, composed of cells of new growth, the inner part of which degenerates later and enters into the formation of the cheesy central mass. Within this tubercle is the young parasite. In later stages this tubercle enlarges until it becomes 3''' in diameter. In this stage the soft interior mass will be firmer. The parasite is always found between the interior mass and the capsule, and is surrounded by the soft, freshly-formed greenish material, which it seems to produce by the irritation of the adjacent capsule. When the parasite attains its adult size it evidently breaks from the tubercle and thereafter lives in the adjacent bronchioles. There is quite a difference in the external appearance of the little tubercles during the different stages of growth. In the earlier stages they appear as little blood-red spots just beneath the pleural coat of the lung; later they look like little brownish fluid-filled tumors, surrounded by a red zone; still later a yellowish, green, cheesy material appears in their center, and the tumors present a greenish-gray appearance. The gray is due to the thickened capsule and a thickening of the pleural coat of the lung over the little tubercle. There is usually a slight elevation of the surface of the lung over these nodules, but this feature is dependent on the depth at which the nodule is situated. They may occur at any depth in the lung substance, but are usually near the surface. When the parasite escapes from the nodule a new phase of the disease begins. It wanders through the bronchi until it meets one of the opposite sex, when they mate. Soon after the female begins to lay eggs in the bronchioles and alveoli, which she infests, and these eggs in turn hatch into young worms. These young worms are very lively, and help to increase the disturbance of vital functions of the lung surrounding them. That part of the lung then becomes as if sodden, the air tubes fill with eggs, worms, cast-off epithelial cells, mucus, wandering cells, and air globules; the tissues of the walls of the alveoli become thickened and encroach upon the contents, and the function of the part is entirely suspended. The effect of the worm and its brood at this stage is to produce a pneumonia, hence the disease has been termed *verminous pneumonia*. This pneumonia is limited to the neighborhood of the parasite and does not extend beyond. The patches are from 1 to 2.5ᶜᵐ in width, but in those recently formed they rarely extend more than 2 or 3ᵐᵐ deep. The injury seems to be mainly a mechanical effect, due to the irritation set up by the parasites. When one of these patches is cut into a frothy liquid exudes, bearing quantities of eggs and embryos in all stages of development. They may be seen with a glass magnifying six diameters.

In later stages of the disease the tubercles become little hard masses. These have been said to be calcareous, but they are not soluble in acid, and seem rather to be the contracted, hardened remains of the cheesy mass. There are sometimes found in certain lungs raised patches of a rather dry, emphysematous tissue, which seems to be due to the deeper lying parasites. In other lungs the patches which once showed the pneumonia have becomed thickened, firmer, denser, and a cut across them shows the thickening to extend to a considerable depth.

The abundance of the nodules and patches of pneumonia is very variable. There may be a dozen nodules of different sizes and two or three patches, or the nodules may be diffusely scattered over the whole posterior surfaces of the lung, or there may be associated with them numbers of patches due in part to the close proximity of the nodules and in part to the extension of the disease. In other cases there may be a few of the nodules with a series of patches ranged along the dorsum of the lung. Each lung seems to present a slightly different phase, dependent on the degree and the time of infection and possible reinfection.

Source of infection.—That verminous pneumonia is caused by a worm (*Strongylus ovis-pulmonalis*) and that sheep become infested while feeding or drinking has already been enlarged upon. It is obvious, therefore, that the best way to keep the sheep well is in some way to prevent them from becoming infected with the parasite while feeding.

Preventive treatment.—In giving rules for prevention the value of knowing the complete life history is fully illustrated. The unexplained gaps in this history are two, viz: there has been no complete demonstration of the manner and place in which the parasite spends its life between the time of its escape from the lung of one animal and its reception into that of another, nor has it been demonstrated that the worm must escape from the lung before it may complete its development. This latter item is an important one, for if the worms can continue multiplying indefinitely in the lung then there is little hope of freeing a sheep after it is once infected. On the other hand, if an infected sheep is to be regarded as incapable of continuing the infection within itself then the case is more hopeful. If the parasite must become parasitic on a second host while external to the sheep, as some claim, this is an important factor in its life history, for its continuance then depends on the presence, abundance, and seasonal appearance of this second host, and influences adverse to the life of the second host would be unfavorable to the parasite.

As the parasites are present in the lungs of sheep throughout the year in all stages, this theory does not seem to hold good. The infection of lambs is proof enough that the parasites are continually passing from one sheep to another, and whatever be the mode of living there are certain precautions which may be taken to keep the sheep less infected if not to entirely exclude the worms. The older sheep, which seem to be more infected and which are the source of infection for young ones, should be marketed. Lambs should be weaned as early as they safely can be, separated from the older sheep, pastured in fields where there have been no sheep since the previous winter at least, and never allowed to pasture, water, or yard after infected animals.

Sheep should be supplied with water from running streams or troughs, and should not be allowed to contaminate the water in any way. Filthy drinking water is one of the most prolific sources of the parasite.

There are two kinds of seasons which especially favor the production of parasitic diseases. The one is a very wet, warm season, during which the parasites seem to be able to live on the damp ground. The other is a very dry season, when the pools of water become diminished and stagnant, and whatever parasitic eggs or embryos there are in them are gathered into so small a volume of liquid that sheep drinking of the water become more readily infected. Wet, damp pastures, and pastures with puddles in them are alike favorable to the worm diseases. Sheep should be excluded from such places as much as possible.

A constant watch of the condition of the lungs in dead and slaughtered sheep will enable the flockmaster to judge of the progress that his care in preventing the disease has made.

Medicinal treatment.—There is no medicinal treatment that can be profitably followed. Salting, grain-feeding, and healthful surroundings are required not only to keep up the health of the animal for the production of wool but to fit it for the market, which is the best place for seriously affected sheep.

STRONGYLUS OVIS-PULMONALIS, Diesing.

PLATE XXIX

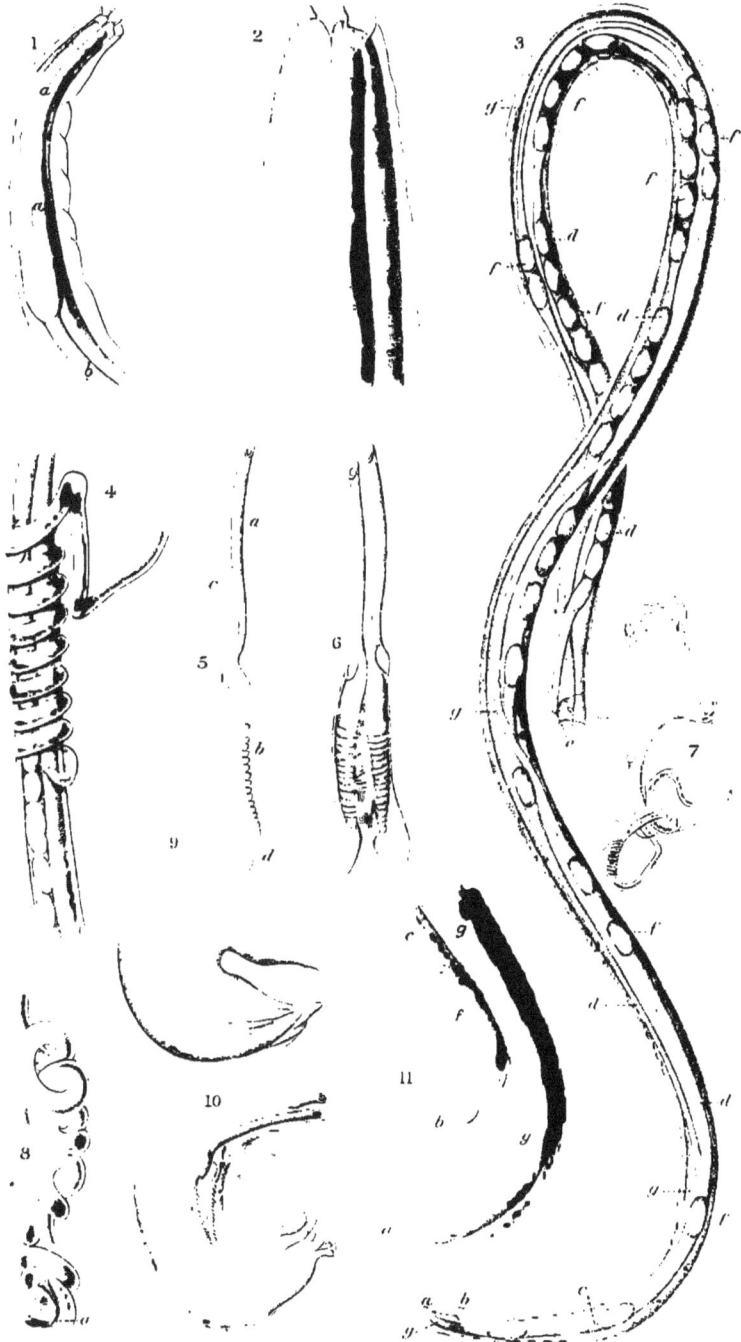

STRONGYLUS OVIS-PULMONALIS
(The Hair Lung Worm.)

STRONGYLUS OVIS-PULMONALIS, Diesing.

PLATE XXX.

Portion of left lung slightly affected by the *strongyli*. The purplish spots are those more recently invaded. The small gray spots are older. The large gray spots are caused by the worms and their young, which have produced an appearance of local pneumonia.

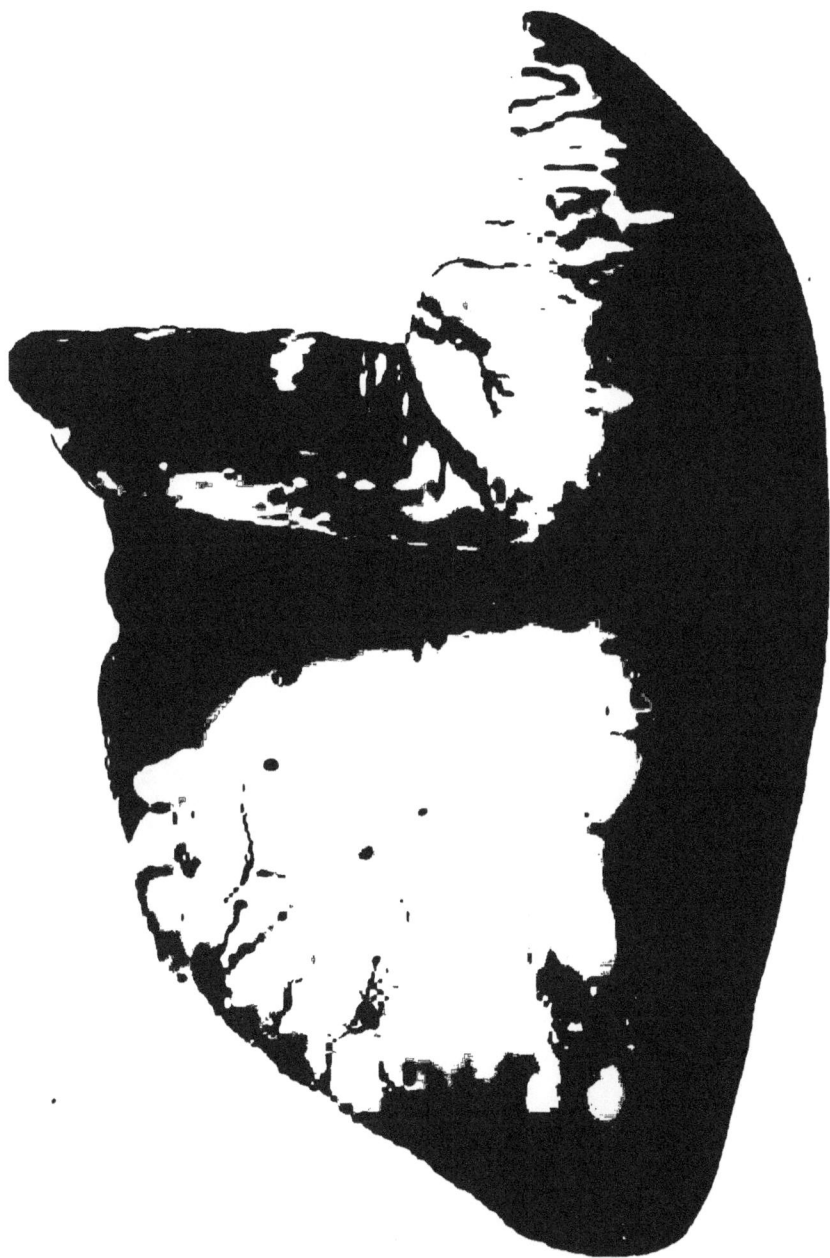

.

STRONGYLUS OVIS-PULMONALIS, Diesing.

PLATE XXXI.

Left lung diseased by *Strongylus ovis-pulmonalis*, the hair lung-worm. Natural size. Each dot is caused by the irritation set up by a young worm, and its size corresponds to the age of the worm. The larger patches consist at first of separate dots; as these enlarge they run together and finally become so fused that their identity is lost. The patches show the stage at which the worms become adult and produce their young, which wander into the adjacent air cavities.

PLATE XXXI

Raines, del.

Amer. S. Co. Lithocaustic

SURFACE OF LUNG DISEASED BY STRONGYLUS OVIS-PULMONALIS.

STRONGYLUS OVIS-PULMONALIS, Diesing.

PLATE XXXII.

Portion of right lung, exhibiting an advanced stage of the hair lung-worm disease. The small dark spots show the youngest stages, the large patches show the disease well advanced, while the large light spots are the oldest. A section cut across one of these shows the depth at which the lung is affected.

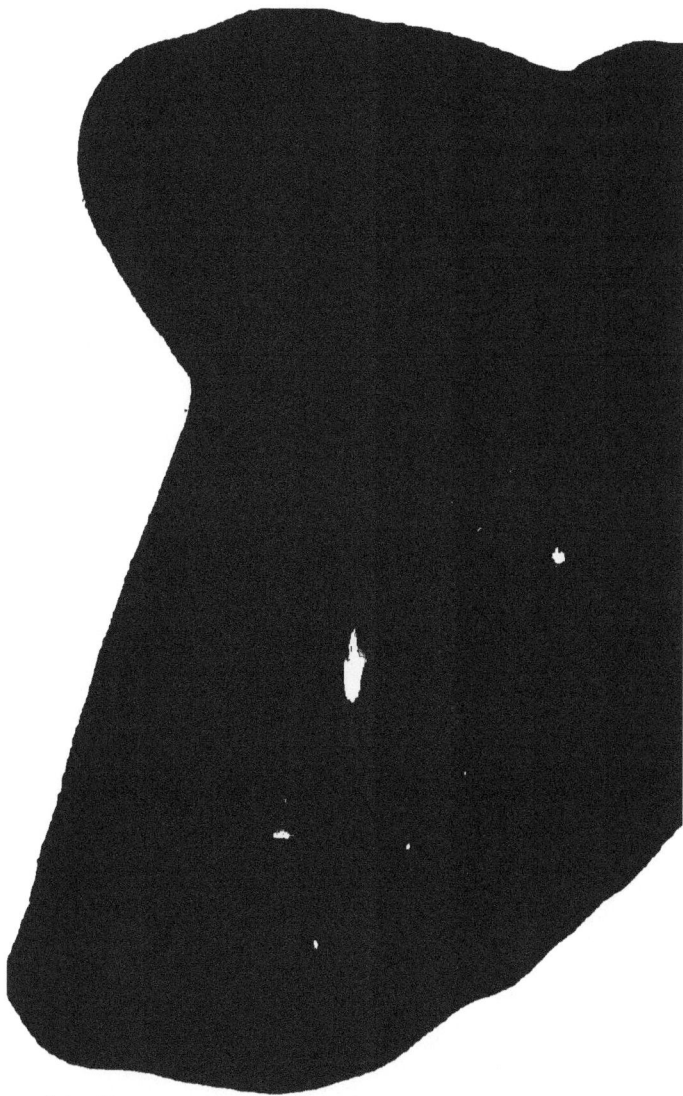

Haines, del.   A Hoe & Co lit

SURFACE OF LUNG DISEASED BY STRONGYLUS OVIS-PULMONALIS.

STRONGYLUS OVIS-PULMONALIS, Diesing.

PLATE XXXIII.

Fig. 1. Section of lung tissue through two small tumors caused by the worms ×20: *a*, caseous degeneration of tissue in the center of the tumor; *b*, the same in the pathway of the moving, growing worm; *c*, cut fragments of the worm (the pathway of the worm is interrupted between *a* and *b*, because the plane of the section did not include it); *d*, a bronchus into which the parasite has almost found its way; *e*, portion of a second tumor made by another worm; *f*, nearly normal tissue.

Fig. 2. Section through an older tumor at the stage which has been likened to pneumonia, ×20: *a*, tumor with fragments of worms; *b*, fragments of an adult worm; *c*, eggs in segmentation stage; *d*, embryos somewhat developed; *e*, young embryos; *f*, bronchi; *g*, nearly normal tissue.

Fig. 3. An enlargement of *b*, Fig. 2, showing fragments of adult worm in the bronchi and alveolæ.

Fig. 4. An enlargement of *e*, Fig. 2, showing young worms in the alveolæ.

Fig. 5. An enlargement of *c*, showing segmenting eggs in alveoli.

Fig. 6. An enlargement of *d*, showing developing embryos in alveoli. In Figs. 5 and 6 the outlines of the egg-shells are not shown. Figs. 3–6, ×90.

(These illustrations were made from specimens selected from a number of serial sections which were stained with alum-carmine; the dots represent the nuclei of the cells. All sections show the great multiplication of cells about the points of irritation, whether excited by the adults or embryos.)

Fig. 7. *a*, embryo of *Strongylus filaria*, and *b* embryo of *S. ovis-pulmonalis*, each equally enlarged to show comparative differences in size and outline.

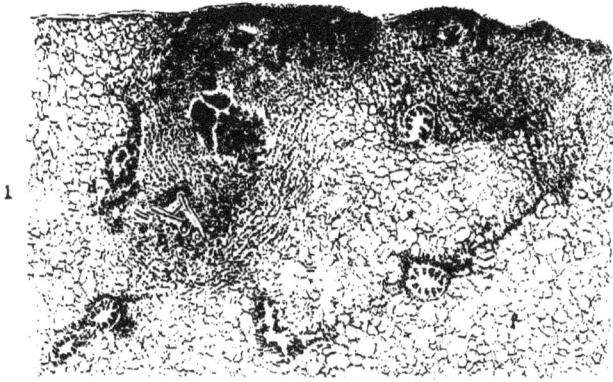

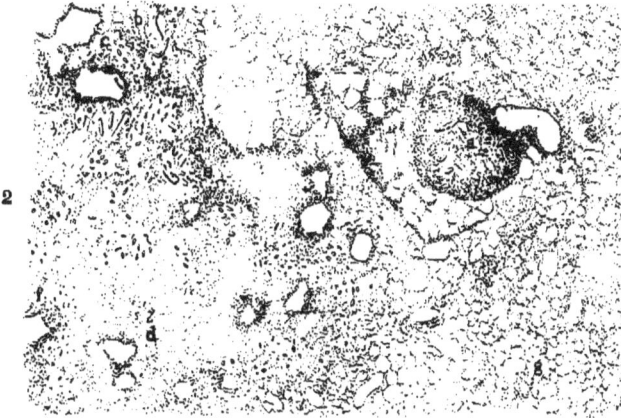

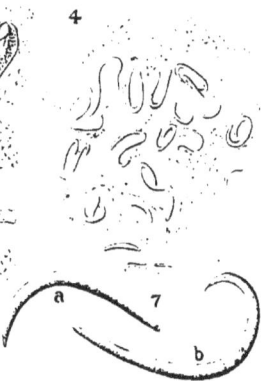

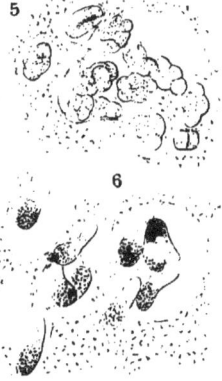

PLATE XXXIII

A Brown & Co. Lith. Melbourne

SECTIONS OF LUNG DISEASED BY STRONGYLUS OVIS-PULMONALIS.

THE THREAD LUNG-WORM—VERMINOUS BRONCHITIS—HUSK OR HOOSE—PAPER SKIN.

STRONGYLUS FILARIA, Rud.

Plates XXXIV, XXXV, XXXVI.

The thread lung worm, or *Strongylus filaria*, is the best known of the sheep lung worms, for the reason that at times it causes extensive epizootics in the flocks, and that the worm is large enough to see when the bronchial tubes are slit and spread open. From personal observation it appears to be much rarer than *Strongylus ovis-pulmonalis*, and the disease it causes much less extensively distributed as to number of animals infected than that produced by the latter. In most of the American literature on this subject the disease caused by the hair lung-worm seems to be ascribed to the thread lung worm, and no mention is made of the former.

Description.—Male, 33 to 54^{mm}; female, 55 to 80^{mm}. Worm filiform, white, with a dark hair line showing throughout its length; head obtuse, without noticeable papillæ or wings; mouth circular, naked; unicellular neck glands quite large; cuticle longitudinally striate. Male: Bursa shallow, campanulate, opening laterally; five sets of costæ; the dorsal are trifid, the lateral bifid, and the ventral separated. Spicula arcuate cylindrical; 3.35^{mm} long by 0.075^{mm} wide; short, very thick, dark brown; chitinous portion a curved fenestrated conical tube; fleshy portion a membrane, which forms a bulb-like expansion toward their free end. Female: Vulva three-sevenths of her length from the head; uteri symmetrically directed anteriorly and posteriorly; posterior oviduct becoming continuous with the uterus near its flexure at the tail; ovo-viviparous; eggs ellipsoid, 0.075 to 0.120^{mm} long; 0.045 to 0.082^{mm} wide. Embryo 0.25 to 3^{mm}.

The life history of *Strongylus filaria* is in general that of other parasites. In some way the young worms arrive in the bronchi, grow, develop, become adult, mate, and lay their eggs in the surrounding mucus. The eggs laid are not true eggs, for each egg-shell contains a young worm within, a feature which is described by calling the female ovoviviparous. The inclosed young escape from the shell, and many of them are expelled from the lungs in the coughing fits along with other discharges. These young, which are scattered about watering-places, pastures, sheep-yards, or corrals serve as infecting and reinfecting material for a considerable length of time. Professor Leuckart (*Entwickelung d. Nematoden, Arch. d. w. Heilkunde*, 1865, p. 299), kept the young of this species alive for several weeks on damp earth, and observed them pass through a stage in which they molted or threw off their skins,

201

after which many died. Baillet (*Colin, G., Bull de l'Acad. de Méd.*, t. XXXI, 1866, p. 874) preserved them alive in water for several months. Ercolani (*Neumann, Traité des Maladies parasitaires*, p. 515) is authority for the statement that they can be resuscitated after being dried a year by putting them in water. The writer has kept them in stagnant water for weeks. Ercolani's statement is by far the most remarkable, and accounts for results obtained in an experiment in which the writer kept sheep for five months on a narrow dry pasture, supplying them with water from a pump only. When these sheep were examined they were found affected with *Strongylus contortus, S. filicollis, S. ventricosus, Dochmius cernuus,* and *Tænia expansa* in very young and old stages. The eggs of these were introduced on the pasture from two or three older sheep which were with the younger ones, or possibly by the young sheep themselves, some of which were between three and four months old at the time. Two of the lot were born and raised under experimental supervision, and these were also infected. *Strongylus filaria* was not present, but it was not discovered in any of the sheep from the same lot killed at the time of selection of the experimental animals, nor has any trace of this parasite been discovered in any of the older ones kept at the Experimental Station.

Professor Raillet details experiments (*Recueil de Méd. Vétérinaire Annexe,* 7 Serie, Tome VI, No. 8, April 30, 1889, p. 173) in which he dried embryos of *Strongylus filaria* under different conditions, and found, after a few failures, that some could be revivified as late as sixty-three hours afterward by placing them in water. His success depended on the condition of the embryo at the time of drying.

It may be accepted, therefore, that the young parasite may retain vitality indefinitely, depending on telluric and atmospheric conditions. From Leuckart's experiment it is to be inferred that though moist earth and damp places are favorable for the life of the young parasites, yet they are liable to molt and then may die from the loss of the older and tougher external skin. From Ercolani's and Raillet's experiments we may infer that the drying of the young parasite suspends its functions, which revive again when the surroundings are suitable, and that the parasite is in this state the most dangerous to sheep.

Preventive treatment.—The foregoing indicates that after a farm is once infected the prevention is not an easy matter, for dry embryos may be scattered everywhere. Although the parasite is more abundant at some seasons than at others, yet it may be found in limited numbers at all seasons, and animals affected will distribute the eggs throughout the year, thus increasing the difficulties of prevention. All animals which show the least appearance of being affected should be separated from the sound ones. The water supplied to the sheep should be pure, *i. e.*, either taken from wells or led into troughs from sources which can not be contaminated. If the sheep are allowed to drink from running water, then all of the brook should be fenced out except where the sheep drink. Dry pastures without bog-holes or sloughs are best for the animals. As the germs live for some time in a dried condition the old pastures should not be used for young sheep at least, nor should the latter be allowed to graze after older sheep which have had the disease during the previous year, nor should the pasture be overstocked so that the grass is eaten to its roots.

Disease.—Verminous bronchitis attacks young animals, those under two years being the more susceptible. Animals poorly nourished and those already weak from other parasitic diseases are also more liable to become a prey to this worn. Damp, warm seasons are most favorable for the preservation of the parasite and the disease it produces. The disease is most prevalent in summer and autumn, becomes less in winter, and disappears in spring time.

The symptoms of this disease, as in verminous pneumonia, are imperceptible in the first stages. It is probable that, beyond the slight but deep cough produced in some of the worst cases, but little else can be noticed. The sheep may have difficulty in breathing when driven or be short-winded. They may be anæmic, as shown by the harsh, dry skin, dry wool, and pale mucous membranes. In later stages the symptoms will be aggravated; difficulty in breathing, coughing, and general debility, associated with an anæmic condition, will be the most prominent symptoms. Occasionally shreddy masses will be coughed up, which, on close examination, will prove to be worms. This is a decisive test of the nature of the disease.

The sheep has a fair appetite, but will gradually lose flesh. In the last stages the bronchial cartarrh is severe, the respiration very feeble and jerky, the cough deep, convulsive, and evidently painful, coming by fits and followed by suffocation, which leaves the patient still more exhausted. The nasal discharge becomes more copious, and contains quantities of embryo and worm fragments. Owing to the diminished respiration productive of anæmia, the skin becomes dry and harsh, and resembles parchment; hence the popular name " paper-skin." The wool is also affected and is easily pulled off, exposing the white, bloodless skin underneath.

Duration.—Death occurs in three or four months either by exhaustion of vital forces or by suffocation. As the first stages pass unnoticed the total time from infection to death is probably nearer five or six months. The duration of the disease depends on the amount of infection, the previous health of the patient, the care it receives, and its vitality. Where the symptoms are very decided the patients rarely survive. The disease is most intense in autumn, and if the sheep do not die, it becomes less intense in winter to more or less completely disappear in spring. When the season has favored the development of the disease and the lambs show severe symptoms, the outlook for their recovery is very unfavorable. A large percentage of those attacked die.

Others fall away in flesh to a serious extent and the growth of the fleece is retarded.

Occurrence.—It is a usual thing to find lungs affected with *Strongylus ovis-pulmonalis*, and more rare to find them affected with *S. filaria*. When the latter occurs it is ordinarily associated with the former, owing to its abundance, but it is easy to separate the two diseases. In the beginning of the *S. filaria* disease the very posterior tip of the lung is affected, turns dark red or grayish, and has a solid feeling and appearance. From this the disease spreads anteriorly, lobe after lobe of the lung becoming involved as the bronchi choke up. These terminal patches are very sharply separated from the adjacent portion of the lung, which appears normal, except that it may be infected with *S. ovis-pulmonalis*, as indeed may be the part infected by *S. filaria*. The cause of this solidification or hepatization (so called because it becomes solid like liver) is the stoppage of the air tubes by the worms and the débris they produce. When they exclude the air from the part the air cells fill with débris and the part becomes solid. Portions of lobes elsewhere may become involved, but more rarely. The anterior lobes often appear red and solid, but it will generally be noticed that in these the red part is thin and not as spongy and resistant as the lobes in the posterior end. This state is due to the air being driven out of the lobes and the walls coming together, producing a state of collapse (carnification or atelectasis).

The solid lung produced by *S. filaria* is often covered by a thickened whitish membrane, the inflamed serous membrane, which often grows fast to the chest or thoracic walls. After the worms disappear, either having been killed by remedies or from some unknown reason, the healing process begins, and the lamb recovers if not too much weakened.

Treatment of this disease is far more hopeful than that of the pneumonia due to *Strongylus ovis-pulmonalis*. It may be dietetic, preventive, and medicinal. In an essay on this disease Mr. Stephen Powers (*The American Merino*, O. Judd Co., 1887, p. 283) says:

To sustain the strength and vitality of a sheep already affected is exceedingly difficult, because the appetite is feeble and capricious. The lamb can seldom be induced to eat enough even of the most nutritious food, to make any considerable impression on it in the way of betterment; and the danger in giving it by force stimulating gruels, etc., is that, owing to its bloodless condition, the process of digestion will be so illy performed that the food will do it more harm than good by causing scours. High feeding is of transcendent importance as a preventive measure; but when the lamb has reached such a pass that vermifuges have to be employed, it is necessary to proceed with great caution in giving rich food.

These remarks commend themselves to all who have had experience with afflicted sheep. Keep the lambs up to the highest point of excellence and health by feeding and they will the better withstand the ravages of the parasites. Corn and oats, bran, chops, and oil-cake are all good fatteners, and should be given in proper proportions. Salt should be placed where the sheep have free access, not only as a diet-

ary article, but for its medicinal influence. In addition they should
have pure, fresh water once or twice a day. When the animals have
become sick good diet should be supplied. As intimated by Mr. Powers,
those animals which seem most in need of food take the least, and if
they do eat it may even be of harm to them. However desirable it may
be to feed animals well as a hygienic measure, still no amount of feeding will keep them from being infected when a season favorable to
the parasite appears. There must therefore be a continual diligence
exercised in keeping the pastures in good condition and the young
sheep especially from becoming infected. As the parasites seem to thrive
best in water, it follows that dry pastures should be preferred. The
danger of infection from pastures should be diminished by limiting the
number of sheep, so that they will not have to eat the grass close to
the roots, and by a judicious distribution of the young sheep on practically virgin pastures. Should a pasture have become permanently infected from long use it should be plowed up and either cultivated a
year or two or allowed to stand idle or surrendered to other stock.
The effect of the cold upon the embryos of these parasites is not yet
known, and it may be that the alternate freezing and thawing which
they sustain is in the Northern States the cause of the destruction of
large numbers of them. Leuckart's experiment of keeping the worms
in moist earth, during which time many molted and died, indicate
that a pasture would be much safer when thoroughly dried after a prolonged rain than before, and also that such a wet time would be more
dangerous for the sheep. A judicious selection of pasturage throughout the year, together with a shifting of the sheep from pasture to pasture as the season and ages of the sheep seem to require, is the best
that can be counseled at present.

Medicinal treatment may be productive of much good, but is usually
resorted to so late that its best effects are lost. Medicines have been
administered with the food by drenching, by fumigations, and by
tracheal injections. Salt and copperas in proportions of from 1 of copperas to 25 of salt, and of 1 of copperas to 4 of salt, the last mixture
being given in wet weather, has been advised (*The American Merino*, by Powers, 1887, p. 285). The weaker mixtures may be kept constantly before the lambs for eighteen months. The stronger should be
alternated every two or three weeks with clear salt. Powers kept it
constantly before the lambs until after the second summer. I would
deprecate the use of copperas for any continued length of time, for it
not only harms the teeth, but if persisted in loses its force as a tonic
remedy. In administering dry medicines in food much of their force is
lost, for they are very apt to accumulate in the paunch or first stomach.
Medicines given by drenching are more expensive in the dosing but
more effective, for small quantities of fluids pass directly into the manifolds or third stomach, and thence into the fourth stomach, especially
if the sheep be thirsty. But few of the many remedies advised are in

the least effective except they be general tonics and stimulants. Many advise the use of anthelmintics, but these are of value only in driving off the intestinal parasites. Turpentine seems to be an exception to this rule, as some of it is eliminated by the lungs and so reaches the worms. Powers (*op. cit.*, p. 283) advises turpentine and linseed oil mixed in equal parts, a tablespoonful at a dose. Mr. W. G. Berry saturates lumps of salt with turpentine, then crushes the salt, mixes with bran, and feeds as a preventive.

Neumann (*Maladies Parasitaires*, p. 517) states that the following have been recommended : Picrate of potash, from 3 to 6 grains per dose, dissolved in oatmeal, water, or mucilage; a mixture of equal parts of turpentine and spirits of camphor, a teaspoonful daily in mucilaginous drink; a mixture of creosote 120 parts, alcohol 500 parts, water 700 parts, dose a teaspoonful; creosote 60 parts, benzine 300 parts, water 2,000 parts, dose a teaspoonful for each patient daily for eight days· Hall (*Veterinarian*, 1868) says that he employed with success 10 drops prussic acid (to be diluted in water) for a dose morning and evening. Neumann adds, however, that experience shows there is little reliance on these methods of treatment, and the administration is, besides, more or less difficult.

The same author states that success is less uncertain with fumigations which penetrate directly to the worms, benumbs them, and provokes a cough by which they are brought up and ejected. The sheep to be treated should be driven into as nearly an air-tight shed or stable as is practicable. Then rags, horns, feathers, hair, old leather, tar, asafetida, etc., should be placed on a red-hot shovel or in an iron pot filled with burning coals or in a tinner's fire-pot. The intensity, duration, and number of fumigations should be graduated according to the tolerance of the sheep. Either some person should subject themselves to the same fumigations, or a very close watch should be maintained in order to prevent the lambs suffocating.

Tracheal injections.—The method of treatment by tracheal injections promises much better results, but should only be practiced by a reliable veterinarian, who can oversee the results and take all necessary precautions. The method has been detailed in the Second Annual Report of the Bureau of Animal Industry, 1885, page 284. It consists of introducing remedies directly into the trachea by means of a hypodermic syringe which cause the death of the parasites. The medicines thus introduced have an opportunity of acting upon the parasites directly, before they are all absorbed by the mucous membrane of the air passages. There is no reason to doubt that they may have even a secondary effect after their absorption if they are naturally thrown off by the mucous membrane of the air-passages and the epithelium of the alveoli, which is the case with most volatile substances.

The method of tracheal injections was first tried by Gobier in the early part of the present century, after learning experimentally that considerable quantities of liquid can be introduced into the trachea without producing suffocation. Delafond some years after conducted some experiments to determine the absorptive power of the air passages. He found that mucilaginous decoctions and solutions of sugar or honey are speedily absorbed when injected into the trachea, inducing slight symptoms of suffocation for one or two hours. He also found that solutions of narcotic agents and stimulants manifest their physiological effects very soon after injection, and that oils and oily medicines produce a congestion of the lungs which is but slowly dissipated, and that even very dilute solutions of mineral and vegetable acids produce inflammation, with copious secretion of mucus, giving rise to symptoms of asphyxia and even leading to death.

Dr. Levi, of the University of Pisa, has recently applied this method in the treatment of a number of diseases (*Manuel pratique des injections trachéales dans le cheval,* 1883). His experiments also tended to show that the mucous membrane absorbs very rapidly, and is therefore less apt to suffer from the injection of irritating substances than if the absorption were less rapid. He also determined that the injection of small quantities of oily substances is not dangerous, the oil probably being emulsified and absorbed. Finally, there is always a slight reduction in the number of respirations, amounting to about three or four per minute, after the introduction of liquids, even when distilled water only is injected.

Without entering into interesting questions concerning the administration of medicines in this manner in other diseases, which are discussed at length in the work mentioned, we find that the author has experimented on but one case of lung worms to test the efficacy of the method. Others, however, have reported cases in which their success justifies a detailed account of the method for future application.

The instrument to be used is a simple hypodermic syringe holding from 1 to 2 fluid drams. The needle of the syringe must be provided with a removable solid rod or trocar, so as not to become plugged when it is pushed through the skin and walls of the trachea. As the needles are apt to break, a number of them should be kept on hand. After the operation the syringe should be carefully washed in pure water, the piston supplied with a drop of olive oil, and the trocar replaced in the needle.

It is best to disinfect by filling the syringe and needle with a 5 per cent. solution of carbolic acid, or a 0.1 per cent. solution of mercuric chloride* before washing in pure water. The disinfection, however, is not absolutely necessary in this operation if the syringe and needle be kept thoroughly clean.

To administer the medicine first fill the syringe and place at the side. Hold the sheep for drenching, and extend the head of the animal so as to fix and make prominent the trachea, which will be felt as a tense elastic tube along the middle line of the neck. The most convenient point for the introduction of the needle is at about the middle of the length of the neck. It must be remembered that some care is to be observed, as the trachea is near some important structures on either side—the jugular vein, the carotid artery, and the pneumogastric nerve. Having fixed the trachea with the left hand, the needle with the trocar is inserted beneath the skin, and then an interannular space is sought so as not to pierce a cartilaginous ring. Or the needle may be pushed directly into the trachea without necessarily avoiding a cartilaginous ring. The unimpeded movement of the free end of the needle as if in an empty space is a sure sign that the needle is in its proper place. The trocar is now removed, the syringe screwed upon the needle, and the contents very slowly forced into the trachea. Before the needle is finally withdrawn Dr. Levi thinks best to wash it out with some pure water so as to remove the injecting fluid. In withdrawing the needle this might accidentally be discharged in the wound made by the needle and set up inflammation if the substances introduced be irritating. How this washing out is to be done he does not state. It seems that a small pipette or medicine-

---

* The former is prepared by adding 5 parts by weight of pure carbolic acid to 100 parts by weight of pure water previously heated; the latter by adding 1 part of the corrosive sublimate (a violent poison) to 1,000 parts of water.

dropper filled with water and inserted into the end of the needle would suffice to wash it out, or drawing back the piston of the syringe would leave the needle comparatively empty. The needle might also be washed out by removing the syringe, washing it out, filling with water, and forcing a few drops into the trachea through this needle. This, however, would cause unnecessary delay before the animal is released, and is therefore not to be recommended. The simplest method, then, to empty the needle would be to draw back the piston, for the discharge of anything but the purest water into the wound may produce more irritation than the medicinal substances themselves. The animal should be watched for some time, especially after the first operation, to observe how the injection has been borne, and whether any symptoms arise which indicate difficulty of breathing.

If, as has been suggested, a slight incision be made in the skin before introducing the needle, and if a cartilaginous ring be avoided in piercing the trachea, the ordinary needle with beveled extremity will be sufficient, and the trocar may be dispensed with. When the needle has entered the trachea, a slight hissing noise, due to the entrance and exit of air with each inspiration and expiration, indicates that the needle has reached its destination and is not plugged.

The substances to be injected should have distinctly vermicide properties, without being at the same time too irritating or poisonous in their effects on the animal. Levi gives two formulæ which he used with success upon a sheep. The worms were discharged in three days and the catarrh cured : Iodine, 2 parts ; iodide of potash, 10 parts ; distilled water, 100 parts, by weight.

Begin with half a dram of this solution, add half a dram of water, and increase by half a dram of the above solution each day up to 5 drams. Another remedy is the following : Mix equal parts of turpentine and olive oil, and inject from 1 to 4 drams. In this case the writer probably intended to state that the dose should be increased from 1 to 4 drams on successive days.

Eloire (*Recueil de Med. Vet.*, 1883, p. 683) gives the following formula : Ordinary oil of poppy and oil of turpentine, each 100 parts ; carbolic acid and purified oil of cade, each 2 parts.

The oil of poppy, being a bland oil, does not possess any medicinal properties and may be replaced by olive oil. Each sheep to receive about 2 drams a day for three days.

Six animals treated in this way showed immediate improvement and were finally cured. Penhale (*Veterinarian*, 1885, p. 106) reports immediate relief and ultimate cure in two calves by injecting the following mixture : Oil of turpentine, 2 drams ; carbolic acid, 20 drops ; chloroform, ½ dram.

One-half of this amount may be given to a sheep and the dose subsequently increased if necessary.

Hutton (*loc. cit.*, p. 62) reports favorable results in six out of eight cases by injecting the above liquid, in which 1 dram of the tincture of opium was used in place of chloroform.

This completes the list of remedies thus far suggested and tried. The favorable testimony, though not abundant, is very encouraging. There are many substances, no doubt, the use of which might be more beneficial than those mentioned, but nothing can be said of them until they have been tested.

The dose for young sheep should be proportioned to the age and size of the animal. The preparations with turpentine seem to have given the best results. During treatment the patients should receive the best of care.

The prevention of this disease is very desirable, though it may never be completely attained. If a farm is completely free from it in the first place, then prevention simply lies in not allowing infected sheep to be brought on the premises. All purchases of sheep should be from flocks which have shown no signs of the disease in preceding years. Brooks which run from pasture to pasture offer a chance of infection where the neighbors' flocks upstream are infected. Strange sheep should not be pastured unless they are known to be free from parasites. Feeding and care to keep up the general health are essential. Careful separation of affected animals should be practiced, and the worst diseased ones may be slaughtered. Treatment should not be neglected. During treatment it is best to keep the sheep up, and after the course of treatment is concluded they should be turned into new pastures.

23038 A P——14

STRONGYLUS FILARIA, Rud.

PLATE XXXIV.

Fig. 1. Adult female, ×3: a, head; b, vulva.

Fig. 2. Adult male, ×3: a, head; b, bursa and spicula. The dark line in Figs. 1 and 2 is the intestine.

Fig. 3. Cephalic end: a, mouth; b, œsophagus; c, intestine; d, unicellular glands.

Fig. 4. Middle portion of female: a, vulva; b, vagina; c, c, uteri with developing eggs.

Fig. 5. Piece of skin showing striæ.

Fig. 6. Spicula: a, the fenestrated chitinous cylinders; b, the bulb like enlargement of the surrounding membrane.

Fig. 7. Caudal end of female: a, anus; b, b, intestine; c, loop of the caudal uterus; d, ovary.

Fig. 8. Caudal end of male, ventral view: a, intestines; b, seminal reservoir; c, the torn edges of the spread bursa; d, ventral costæ; e, ventro-lateral; f, lateral; g, dorso-lateral; h, dorsal; i, spicula.

Fig. 9. The same lateral view.

Fig. 10. a, female, natural size; b, male, natural size.

Fig. 11. Eggs showing various stages of development of embryo in the following order: a, b, c, d; e, embryo escaped from shell.

PLATE XXXIV

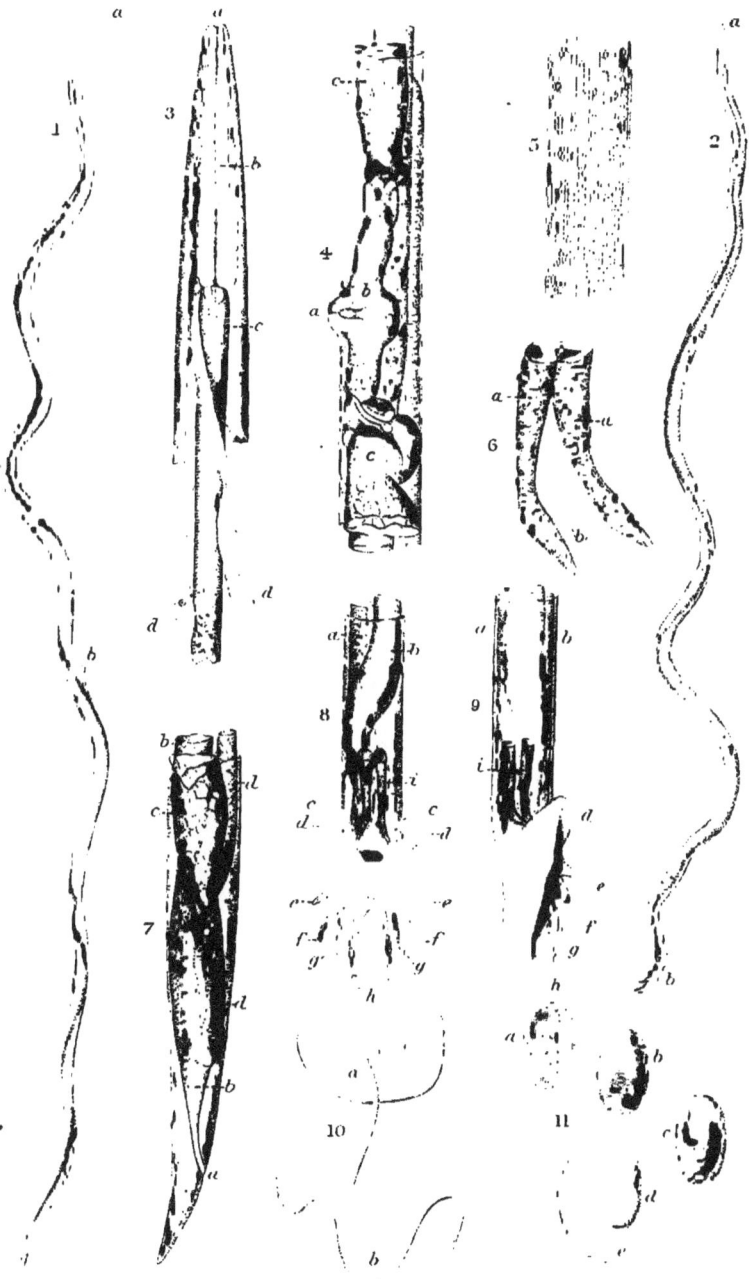

STRONGYLUS FILARIA,
(The Thread Lung Worm.)

STRONGYLUS FILARIA, Rud.

PLATE XXXV.

Portion of right lung of lamb, dorsal view. The affected region is the dark colored posterior end. It is clearly marked off from the healthy portion. The dark red spot about the middle of the figure is also caused by the worms, but it is exceptional to find these lobes affected in early stages.

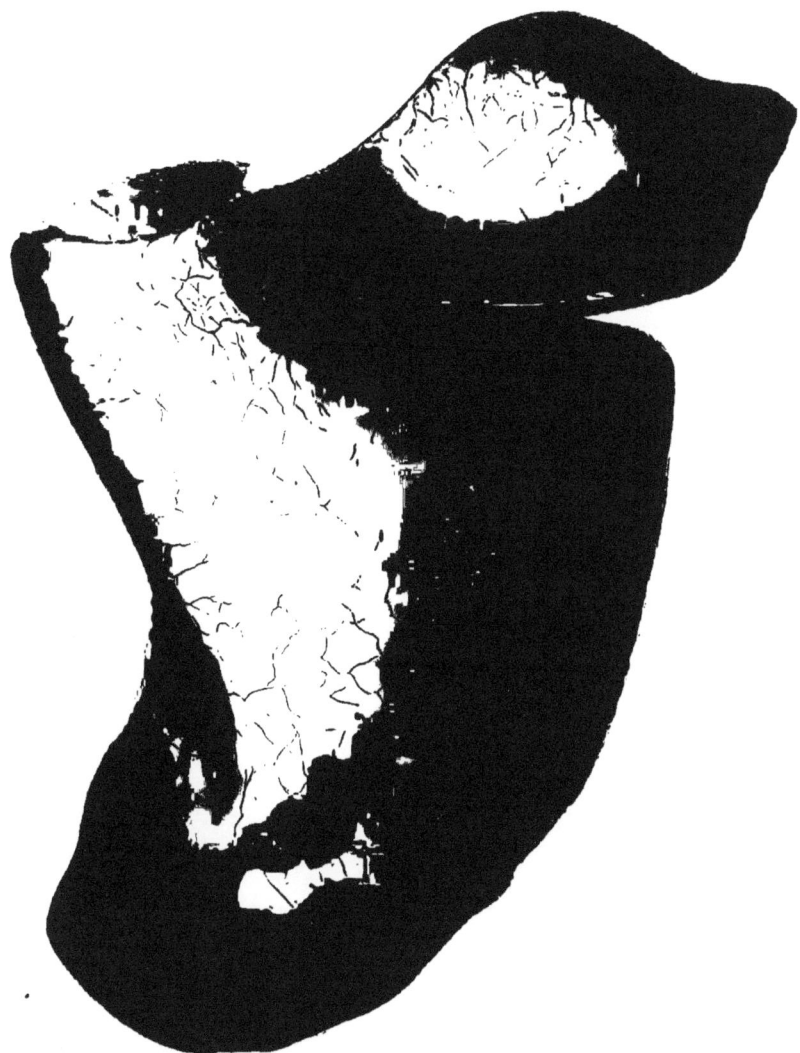

214

STRONGYLUS FILARIA, Rud.

PLATE XXXVI.

Portion of right lung of lamb, ventral view. This is the same lung figured in Plate XXXV. The affected portion is in the posterior end. The lung tissue has been dissected to show the tracheal branches and the bronchi. The bronchi leading to the dark-colored affected region are filled with the life-sized figures of the *Strongylus filaria.*

PLATE XXXVI

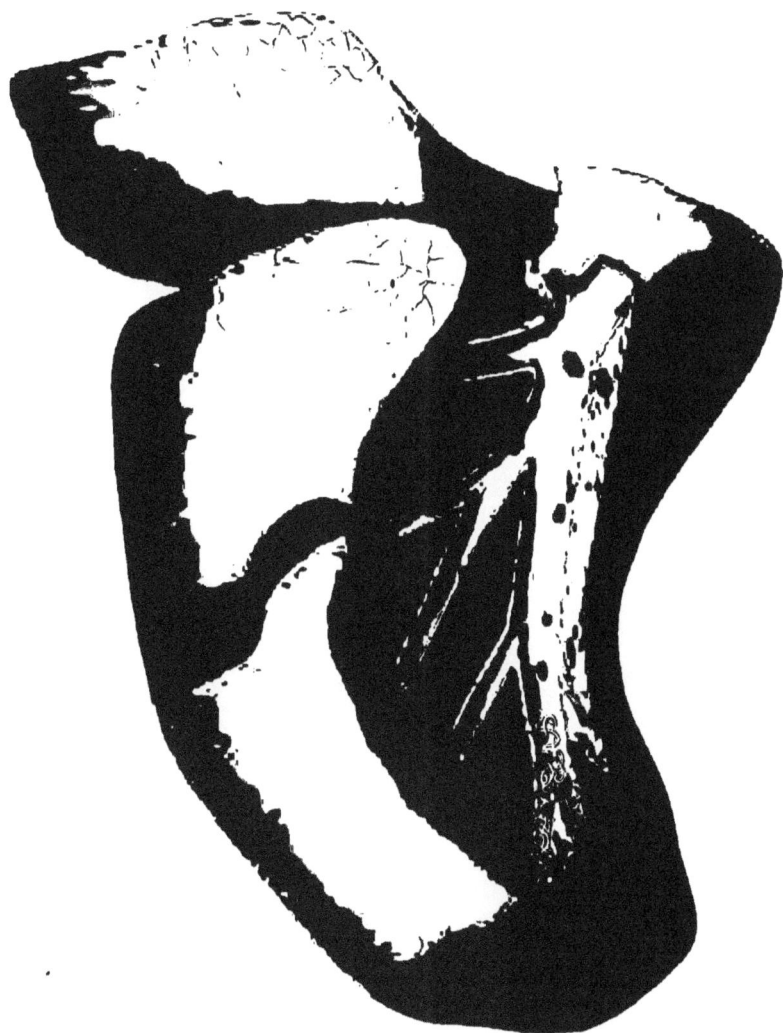

Haines, del.

LAMB'S LUNG INVADED BY STRONGYLUS FILARIA.

INDEX.

●

www.ingramcontent.com/pod-product-compliance
Lightning Source LLC
Chambersburg PA
CBHW021511210326
41599CB00012B/1212